Tensor Properties of Crystals

Tensor Properties of Crystals

2nd Edition

D R Lovett

Department of Physics
University of Essex

CRC Press
Taylor & Francis Group
Boca Raton London New York

CRC Press is an imprint of the
Taylor & Francis Group, an **informa** business

First edition published 1989 by IOP Publishing Ltd.

This edition first published 1999 by IOP Publishing Ltd.

Published 2017 by CRC Press
Taylor & Francis Group
6000 Broken Sound Parkway NW, Suite 300
Boca Raton, FL 33487-2742

© 1989 by Taylor & Francis Group, LLC
CRC Press is an imprint of Taylor & Francis Group, an Informa business

No claim to original U.S. Government works

ISBN-13: 978-0-7503-0625-6 (hbk)
ISBN-13: 978-0-7503-0626-3 (pbk)

Visit the Taylor & Francis Web site at
http://www.taylorandfrancis.com

and the CRC Press Web site at
http://www.crcpress.com

British Library Cataloguing-in-Publication Data

A catalogue record for this book is available from the British Library.

Library of Congress Cataloging-in-Publication Data are available

Typeset in the UK by KEYTEC, Bridport, Dorset

Contents

Preface

In the physical sciences, properties of materials are of considerable importance. Most of the technological applications of science depend on an understanding of these properties. Crystalline solids are perhaps the most important single type of material and, increasingly, single crystals are of especial significance. This arises partly because modern preparation techniques make their growth easier and hence there is greater availability of single crystals than previously; but also the high purity and reproducibility of good single crystals are essential for high-quality solid state devices.

Crystals consist of regular arrangements of atoms, and as a direct consequence of this patterning the physical properties of crystals may differ in different directions. A physical quantity such as electrical conductivity may take on a continuous range of values depending on the direction in which the quantity is measured within the crystal. Nevertheless, there will be a restricted number of independent coefficients necessary to evaluate the conductivity in any direction. To discuss which independent coefficients are required for any type of crystal requires a basic understanding of crystal symmetry. To then use the magnitudes of these independent coefficients to calculate the conductivity in any arbitrary direction requires the application of mathematical quantities called tensors. Because it is possible to explain the way in which these properties vary in different directions by the transformation of tensors, we often call them tensor properties of crystals.

Hence this book will start with a discussion of crystal symmetry. Then we will talk about tensors and in particular the mathematical notation of tensors. Once the basics have been established, the application of tensors to a wide range of physical properties will be studied. These applications will be introduced in an increasing order of complexity. Hence, if time does not allow for a study of the complete range of topics, it should be possible to leave out the later chapters of the book without losing a basic understanding of the subject.

This book arose from a series of lectures given as part of a solid state physics course to second-year physics undergraduates at the University of Essex. I would like to acknowledge the interest and help I have received from colleagues and students during the preparation of the lecture course and the book. In particular, I am grateful to Gillian Mander for reading the book and making helpful comments.

David Lovett
Colchester, Essex

Preface to Second Edition

This textbook has been well received and has already gone through a number of reprintings. With the reprintings, minor corrections were included and I am very grateful to those who suggested certain changes. In producing this new edition it has been felt important not to change the basic format of the book, and early chapters are largely unchanged. However, with the increasing importance of non-linear optics it has been decided to move the sections on optoelectronics to a separate chapter and to expand these to include more on optoelectronic devices. Meanwhile, a short discussion of incommensurate modulated structures is included in the final chapter in view of their relevance to high temperature superconductors and to ferroelectric and ferromagnetic materials.

David Lovett
Colchester, Essex

List of main symbols

This book looks at a wide range of applications of tensors and most of these applications have their own recognised symbols which often overlap. It has been inevitable that many symbols have more than one use.

a, b, c	lengths of sides of crystal cell
a_{ij}	generalised Hall coefficient
c	velocity of light *in vacuo*
c (usually c_{ijkl})	elastic stiffness, Young's modulus
$\mathrm{d}c/\mathrm{d}r$	concentration gradient
C	c-face face centring in Bravais lattice
d (usually d_{ijk})	piezoelectric modulus, also second-order non-linear coefficient
D (usually D_{ij})	diffusion coefficient
D	electric displacement
e	electron charge
e (usually e_{ij})	strain (including rotation)
E	electric field
E_{H}	Hall electric field
F	face-centred Bravais lattice
g (usually g_{ij})	gyration coefficient
G	optical activity constant
h (usually h_i)	density of heat flow
h, k, l	Miller indices (cubic system)
h, k, i, l	Miller indices (hexagonal system)
I	body-centred Bravais lattice
I	total electric current
J	electric current density

k	wavevector
k (usually k_{ij})	thermal conductivity
K (usually K_{ij})	dielectric constant
l (usually l_i or l_{ip})	direction cosines
m	mirror plane
n	refractive index, electron concentration
n_1, n_2, n_3	refractive indices referred to principal axes
n_E	extraordinary refractive index
n_O	ordinary refractive index
n_r	refractive index (right-handed circularly polarised light)
n_l	refractive index (left-handed circularly polarised light)
p	pressure
p (usually p_{ij})	pyroelectric coefficient
P	electric dipole moment
Q	heat energy
r (usually r_{ij})	thermal resistivity
r_{ijk}	linear electro-optic tensor components (Pockels coefficients)
R_H	Hall coefficient
s (usually s_{ijkl})	elastic stiffness, also Kerr coefficients
S_{ij}, S_{pq}	quantities transforming like second-rank tensors
t	time
T	temperature
$T_{ij..}, T_{pq..}$	tensors
u	extension in length
V	voltage, electric potential difference
W	strain energy
x, y, z	Cartesian coordinate system (usually crystallographic or optical)
x_1, x_2, x_3	Cartesian coordinate system (tensors)
α, β, γ	angles defining crystal cell (α between b and c, etc)
χ_{ijkl}	third-order non-linear susceptibility coefficient
ε_0	permittivity of free space
ε (usually ε_{ij})	permittivity components within a crystal
ε (usually ε_{ij} or ε_m)	strain

η (usually η_{ij})	optical impermeability
λ	wavelength
σ (usually σ_{ij} or σ_m)	stress
Π_{ijkl}	strain-optic tensor
ρ	optical rotatory power
ρ (usually ρ_{ij})	electrical resistivity
ρ_{ijk}	Hall components
ρ_{ijkl}	magnetoresistance components
ω	angular frequency

Brackets around crystallographic indices:

( )	lattice planes
{ }	family of lattice planes
[ ]	lattice direction
⟨ ⟩	family of lattice directions

1 Crystals and Crystal Symmetry

1.1 Structure of Solids

Solids are built up from arrangements of atoms. In crystals these arrangements must show regularity and there must be long-range order such that the pattern of the atoms repeats regularly throughout the crystal. Many atoms, particularly metallic ones, can be considered as hard spheres which stack together as closely as possible. If the atoms are all of the same type, and therefore of the same size, they will fit together in a close-packing configuration. Even in this very simple and specific case, it turns out that there are two differently close packed but totally regular patterns. If atoms of more than one type are involved, then the consequence of different sizes (and also of different magnitudes of interatomic forces) can lead to more complicated juxtapositioning of the atoms, and to less spatial filling by the atoms. The overall properties of these crystals will depend on the properties of the constituent atoms. But the properties are likely to vary in different directions because the atomic patterning will differ slightly in different directions.

1.2 Close-packing Structures

It is easy to see in two dimensions that equisized balls pack together so that each is surrounded by six others as shown in figure 1.1(*a*). When one comes to place the next layer on top there is a decision to be made; one places a ball at either interstice X or at interstice Y, but not both. If one then goes on to complete the layer, there will be nothing geometrically different between the two cases. We call the lower layer A and the upper layer B (i.e. figure 1.1(*b*)). For the third layer one again has a decision to make about interstices: one can start with interstice Z which creates a third layer of atoms (C) whose centres are vertically above the centres of the atoms in the first layer (i.e. layer C is equivalent to the layer A as shown in figure 1.1(*c*)); or one can start with the interstice above Y (or the one above X depending on the original choice of interstice and hence which interstice is now available). By doing this we create a third layer which sits vertically above neither layer A nor B. It is a new layer C as shown in figure 1.1(*d*). One needs to go to the fourth layer before one positions a layer vertically above the first. We can then continue with a regular pattern of layers. In the first case we go on with an arrangement AB, AB, AB, etc, and in the second case with ABC, ABC, ABC.

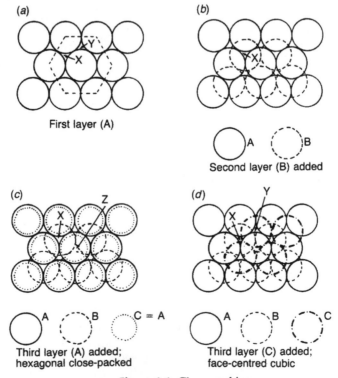

Figure 1.1 Close packing.

These two very simple arrangements illustrate a number of features used in describing crystal structures and in explaining their properties. Even though the packing around any single atom is the same, the packing for the two cases is different once one gets beyond the nearest neighbour. Also, although when we look down onto the paper both show hexagonal symmetry, the ABC arrangement is described as face-centred cubic and a cubic cell is defined to described this structure. Figure 1.2 shows the cubic structure with the close-packed layers clearly marked. One can also see in this figure the threefold symmetry exhibited by a cube when one looks from one corner to the diagonally opposite corner. What is meant by this statement is that when one rotates the cube by 120° it looks the same. If one does this three times one returns the cube to its original position. It is not always realised that threefold symmetry is the fundamental symmetry attached by crystallographers to all those crystal structures which they classify as cubic.

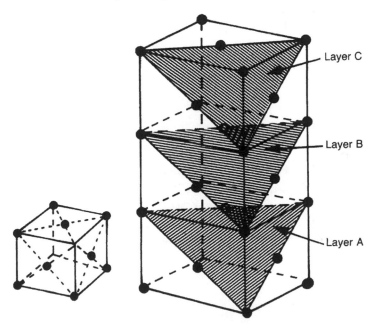

Figure 1.2 The face-centred cubic structure.

The AB (AB. . .) structure is classified as hexagonal as is clearly shown in figure 1.3. The actual cell which is used is not the hexagonally shaped cell but a cell of one third size, an oblique prism in which the oblique angle is 60°.

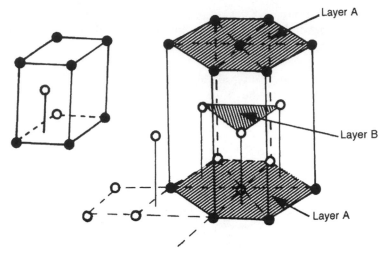

Figure 1.3 The hexagonal close-packed structure.

So we see from this very simple packing two different cells used as building blocks. The cube contains four atoms per cubic cell because the eight corner atoms are shared between eight cells and the six atoms at the face centrings are each shared between two cells. The hexagonal cell (i.e. the oblique prism) contains two atoms because the eight corner atoms are again shared between eight cells and there is one atom inside the cell. This difference in description highlights the packing variation ABC. . . and AB. . . but at the same time hides much of the similarity. In fact, as the atoms are all the same, one should be able to find a basic cell which contains only one atom. One can choose a parallelepiped which does this. Start from any atom and define the sides by going out in three directions to nearest neighbours. This will work for both cases. Any crystal can be defined by parallelepipeds, and in some structures which show a considerable lack of symmetry this is necessary. However, when we can find a suitable cubic cell this is highly advantageous, particularly as then we are able to use orthogonal axes for describing the directional properties.

So far we have assumed one type of atom. Even if we have different types of atoms in the structure, we still describe the structure in terms of the periodic lattice. The outline of the units (such as the cube and the 60° hexagonal-type cell as discussed above) form the so-called *lattice*. This possesses the translational symmetry. Into this lattice we place the groups of atoms called the basis. The basis will also have symmetry associated with it. It is the symmetry of *both* the lattice and the basis which is important and there are certain symmetries of the basis which go together with a specific lattice. Because it is simpler, we shall start by considering two-dimensional lattices. Examples of these are commonly found within the patterning of wallpaper.

1.3 Two-dimensional Lattices

There are five lattices used in two dimensions and they are called Bravais lattices. The square and the rectangular lattices (see figure 1.4) are self-evident. The hexagonal lattice consists of a 60° equisided parallelogram; three such parallelograms form a hexagon when suitably grouped. The oblique cell is the more general lattice used when the patterning does not fit the alternative lattice shapes. The fifth Bravais lattice is the centred rectangle, and has two lattice points in the cell, a corner point and a point at the centre of the rectangle. Essentially, it is a special case of the oblique lattice. One can obtain an oblique cell of half the area and with one lattice point per cell instead of the two of the centred rectangular cell by taking a cell with corner angle given by $\tan^{-1}(a/b)$, where a and b are the lengths

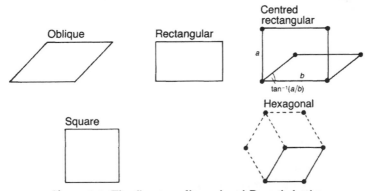

Figure 1.4 The five two-dimensional Bravais lattices.

of the two sides of the rectangular cell. The five lattices constitute four crystal systems as the two rectangular lattices are included in one crystal system.

To complete the two-dimensional pattern, the basis consisting of an arrangement of atoms (or shapes in the case of wallpaper) is added at the position of each lattice point (figure 1.5). The basis may have its own symmetry, and this is called the point group symmetry. The basis may have mirror symmetry and/or rotational symmetry. In the case of mirror symmetry, there is a reflection line through the pattern such that it would appear to reflect across the line onto itself. Rotational symmetry can be twofold, threefold, fourfold or sixfold. But fivefold symmetry and symmetries greater than sixfold cannot be used as they cannot be incorporated into a repeating two-dimensional pattern.

Space lattice   Basis   Crystal structure

Figure 1.5 Lattice plus basis gives crystal structure.

Combining rotational symmetries with mirror symmetries gives 10 point groups which are compatible with the crystal systems as follows:

Crystal system	Compatible point group
Oblique	1, 2
Rectangular	1m, 2m
Square	4, 4m
Hexagonal	3, 3m, 6, 6m

The description 1 means that there is no symmetry associated with the crystal system; the system or pattern only looks the same when it is rotated a full 360° back into its original orientation. Where a two-, four- or sixfold rotational symmetry is associated with a mirror plane, further mirror planes occur as a direct consequence. We can represent the symmetry of the 10 point groups by patterns of arrows. These are shown in figure 1.6.

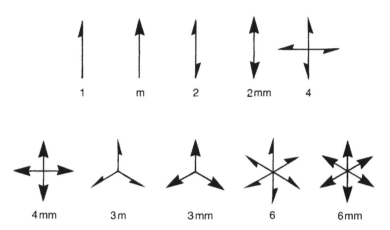

Figure 1.6 The 10 two-dimensional point groups.

A further operation in two dimensions is the glide operation which can only arise in the complete lattice. It consists of a reflection (i.e. a mirror line) plus a translation by half a lattice vector (i.e. by half the length of a side of the unit cell) (see figure 1.7). Two consecutive glide operations return the basis to its original position and orientation in the adjacent cell. On a diagram, a glide line is represented by a thick broken line, whereas a mirror line is represented by a thick continuous line.

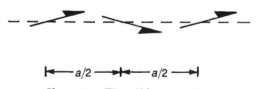

Figure 1.7 The glide operation.

1.4 Three-dimensional Lattices

Extending to three dimensions, we find that there are seven crystal systems which act as building blocks for the complete crystal lattice. The cubic system consisting of a cell of sides of equal length, orthogonal to each other, is the three-dimensional equivalent of the square. Stretching the cube in one direction gives the tetragonal cell with sides of lengths a, a and c. Making all three sides of unequal length but still orthogonal to each other produces the orthorhombic cell. The hexagonal cell (an example of which we have seen already to describe a form of close packing) is analogous to the similarly named two-dimensional cell but has height c. For the remaining cells, it is necessary to define the angles α lying between sides b and c, β lying between a and c and γ lying between a and b. The three remaining crystal systems are monoclinic, with $\alpha = \gamma = 90°$ and $\beta > 90°$, triclinic, which has no relationships between the lengths of the three sides or the three angles and is a general paral- lelepiped, and trigonal. In the trigonal system either a rhom- bohedral cell or a hexagonal cell may be used. The rhom- bohedral cell is smaller and has $a = b = c$ and $\alpha = \beta = \gamma$ all less than 120° and not equal to 90°. The alternative hexagonal cell is three times the size and is usually easier to use.

From the seven crystal systems one obtains the 14 distinct Bravais lattices which exist in three dimensions. There are no other space lattices possible. Each of the constituent points of the space lattices have similar environments. The lattice types are differentiated from each other by the symmetry of the arrangements of their lattice points.

Figure 1.8 shows these 14 Bravais lattices. The symbol P is used for a primitive cell such as for the basic cubic cell with atoms only at the corners. I is used for body-centred cells, F for face centring on each face (of a cube), R for rhombohedral and C for face centring on one pair of opposite faces (called c faces). It is not possible to define a Bravais lattice by saying 'a Bravais lattice is . . .'. The lattices are as shown. Other con- figurations always consist of arrangements which can be shown to fit one of the lattices already listed.

Just as we considered symmetry elements in two dimensions to obtain the 10 two-dimensional point groups, so we consider symmetries in three dimensions and as a consequence obtain 32 point groups (also called crystal classes) in three dimensions. This classification of crystals into point groups is very important when considering the tensor properties.

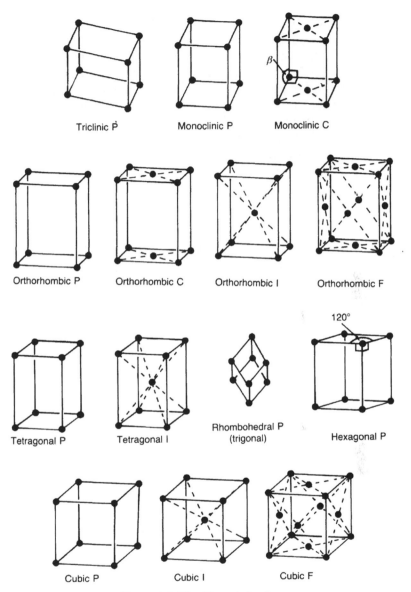

Figure 1.8 The Bravais lattices.

We have mirror planes and rotation axes just as we did for two dimensions. In addition, we have inversion through centres of symmetry and these can be taken in conjunction with rotation axes. Thus a fourfold inversion, which is denoted $\bar{4}$, consists of a 90° rotation followed by an inversion through the centre of symmetry. In figure 1.9(a) we rotate the triangle which represents the atomic patterning (the basis) 90° around the rotation axis and then invert. In figure 1.9(b) we do the same but rotate by 180° to obtain $\bar{2}$. However, we see that the result for this second case is equivalent to the effect of a mirror plane orthogonal to the axis of rotation. Hence the symbol $\bar{2}(\equiv m)$ is not usually used. Other inversion axes are possible including inversion with no rotation, i.e. $\bar{1}$.

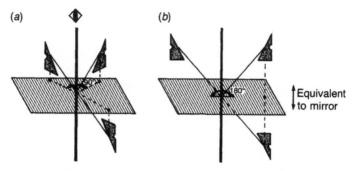

Figure 1.9 Inversion axes: (a) $\bar{4}$ and (b) $\bar{2}$.

Each rotation axis and each inversion axis has its own symbol for use on diagrams:

Rotation axis, n	Symbol	Inversion axis	Symbol
1		$\bar{1}$	○
2	⬖		None
3	▲	$\bar{3}$	△
4	◆	$\bar{4}$	◈
6	⬢	$\bar{6}$	⬡

Rotation is by $2\pi/n$ radians ($360/n$ degrees).

These different symmetries can be combined for different combinations of axes and it can be shown (see for instance McKie and McKie 1974, p.48) that there is a maximum combination of three axes, but not the same three axes for all cases, to give all possible symmetries. As a consequence, we have the 32 point groups shown in table A1 on p.123. In this book we shall be considering only a limited selection of crystal classes and we will go into more detail when necessary. Once the crystal systems and their associated 32 point groups (crystal classes) are put together in a complete lattice, further symmetries are possible. There are glide operations along planes. These operations are equivalent to the glide operations along lines as already described for two dimensions. Such glide planes involve translation plus reflection. There are also screw axes involving rotation plus translation. However, we shall not need to concern ourselves further with these operations other than to note that they involve reflection and rotational symmetries which will affect crystal properties.

1.5 Crystallographic Indices for Planes; Miller Indices

To relate our mathematical notation to real crystals we shall need to define crystallographic planes and directions. For most purposes, it is more useful to define crystal planes in terms of reciprocal lengths. The underlying reason for this is that most physical studies of crystals involve waves, for instance lattice waves, electron wavefunctions, etc, and these are defined by wavevectors which are reciprocal lengths.

Consider a crystal unit cell with cell lengths a, b and c taken in directions x, y and z, with x, y and z usually but not necessarily orthogonal. Consider a plane as shown in figure 1.10(a) such that it intersects the x axis at a, the y axis at $b/2$ and the z axis at $c/3$. We can then define the plane in terms of the reciprocals of intercepts as measured in lattice units. Thus we get (123) and this representation, including the use of the brackets, defines the plane using the so-called Miller indices. If the indices come out as fractions then they must be multiplied by a common numerical factor to bring them to the lowest possible integers. Thus, intercepts of 1/2, 1 and 3 would give reciprocals 2, 1 and 1/3. The plane would be defined in Miller indices as (631). Hence, the Miller index on an axis = 1/(intercept in lattice units) subject to this additional requirement.

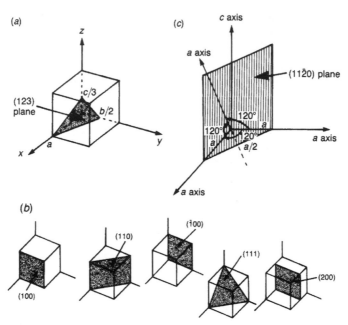

Figure 1.10 (*a*) Defining a crystal plane. (*b*) Miller indices for typical planes in the cubic system. (*c*) Hexagonal system using four axes.

In working out Miller indices, the difficult aspect is to suitably choose the origin. In figure 1.10(*a*) note that the next parallel plane of type (123) would pass through the origin, and the origin should be chosen with this in mind. The indices are used to represent all planes of this orientation with spacings equivalent to the spacing between the plane shown and the next plane through the origin.

Mathematically, the equation representing the set of planes of type (*hkl*) is

$$(hx/a) + (ky/b) + (lz/c) = 1.$$

If any of the intersections with the axes are in the negative direction, the $-$ is placed over the symbol and plane $(\bar{h}\bar{k}\bar{l})$ is equivalent to plane (hkl). However, these two different notations, (hkl) and $(\bar{h}\bar{k}\bar{l})$, are often used to represent opposite parallel faces of the same crystal.

Figure 1.10(b) shows the Miller indices for some typical planes in the cubic system. Planes of similar type such as (100), (010) and (001) in the cubic system can be represented by {100} where { } means that the nomenclature includes all similar planes. For the cubic system this would imply all the possibilities arising from commuting the indices. In the case of the tetragonal system for example, {100} would include (100), (010), ($\bar{1}$00) and (0$\bar{1}$0) but not (001) or (00$\bar{1}$).

One must be careful in the case of the hexagonal system as here four indices can be used to define planes and these are called Miller–Bravais indices. An orthogonal system of axes applied to the hexagonal system hides the $2\pi/3$ symmetry within the basal plane of the cell. Hence, a crystallographic system of indices may be used based on three axes (the 'a' axes) in the basal plane at angles of 120° and a fourth axis (the 'c' axis) orthogonal to these other three (see figure 1.10(c)). A plane in such a system is denoted as $(hkil)$ or sometimes as $(hk.l)$ as $i = -(h + k)$. (This latter equality always applies.) For instance, the plane shown in figure 1.10(c) can be denoted as (11.0) as it intersects the axis along which the symbol h is applicable at a, the axis along which the symbol k is applicable at a also, and the axis along which the symbol l is applicable at infinity. Alternatively, it can be denoted as (11$\bar{2}$0) as it intersects the axis along which symbol i is used at $-a/2$.

1.6 Crystallographic Direction Indices

These are defined in terms of actual distances along the crystallographic axes as measured in crystallographic units. For example if we move a distance from the origin of $2a$ units in the x direction, $3b$ in the y direction and $3c$ in the z direction, the crystallographic direction is [233]. Note we use brackets [ ] to define directions. To include all equivalent directions we write

$\langle 233 \rangle$. It is important to note that the $[hkl]$ direction is orthogonal to the (hkl) plane only in the cubic system. A four-index system for crystallographic directions can be used for hexagonal crystals. This is analogous to the four-index system for planes discussed in the previous section and the four indices measure the number of crystallographic units we move out from the origin along four axes. If we represent the indices by $[uvtw]$, then $t = -(u + v)$.

1.7 Worked Examples on Symmetry

Question 1.A Crystal classes and crystal systems

Ignoring the fine detail, list the symmetries of the capital letters of the alphabet. Put them into crystal classes (i.e. identify the two-dimensional point groups) and crystal systems and hence say how many crystal systems you need to include the complete alphabet. Note that there is not a unique answer to this question as your solution will depend on the style of print face chosen. For instance:

A: vertical m   point group m   rectangular system
                     (crystal system)

Answer

Typically the capital letters can be divided as follows:

Symmetry	Letters	Crystal class	System
mm	HIOX	mm ⎫	
m(horizontal)	BCDE	m ⎬	Rectangular
m(vertical)	T		
2	NSZ	2 ⎫	
1	AFGJKLMPQR	1 ⎬	Oblique

Hence there are four classes and two systems. However, O may have fourfold (or n-fold in effect) symmetry and could be put into a square or a hexagonal system giving a total of five classes and three systems. U etc may have symmetry m and Q could have a mirror line on the slant.

Question 1.B Crystal classes (point groups)

Identify the point groups to be associated with the patterns shown in figure 1.11. (Answers shown in figure 1.12 overleaf.)

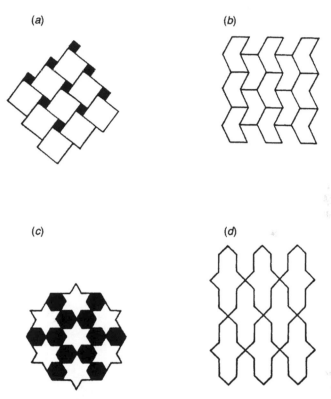

(a)

(b)

(c)

(d)

Figure 1.11 Patterns for Question 1.B.

Question 1.C Lattice planes and lattice directions

(*a*) For the lattice shown in figure 1.13, what are the Miller indices for the listed planes?

(i) BSFT   (ii) HMNP   (iii) OTHO   (iv) JVCW
(v) OKGR   (vi) LTEL.

(*b*) For the same lattice, what are the direction indices for the following directions?

(vii) OT   (viii) LV   (ix) SJ   (x) NK.

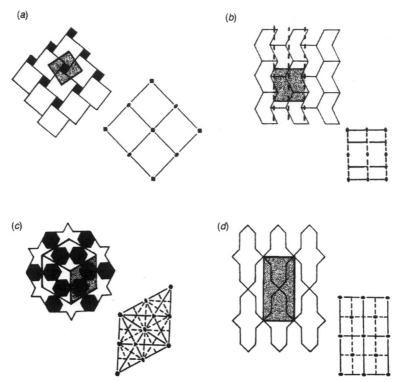

Figure 1.12 Patterns from Question 1.B. with unit cell and symmetry shown. (*a*) 4, (*b*) 2mg, (*c*) 6mm, (*d*) c2mm (i.e. centred rectangular).

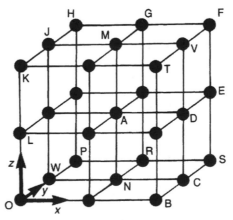

Figure 1.13 Lattice for Question 1.B. and Problem 1.3.

Answer
(a) (i) (100)   (ii) (110)   (iii) (11$\bar{1}$)
(iv) (010), assuming simple cubic   (v) (2$\bar{1}$0)   (vi) ($\bar{1}$12).
(b) (vii) [101]   (viii) [211]   (ix) [$\bar{2}\bar{1}$2]   (x) [$\bar{1}\bar{1}$2].

Problems

1.1 Classify the following Greek and mathematical symbols into crystal classes and crystal systems:

$$\alpha \ \beta \ \delta \ \varepsilon \ \theta \ \Phi \ \Pi \ \infty \ > \int.$$

1.2 Identify the point groups to be associated with the patterns shown in figure 1.14.

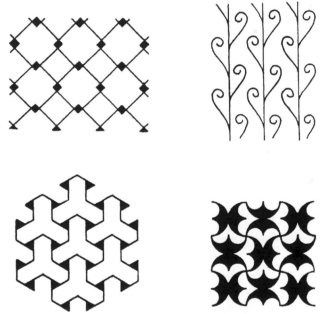

Figure 1.14 Patterns for Problem 1.2.

1.3 (a) For the lattice shown in figure 1.13, what are the Miller indices for the following planes?

(i) KFSO   (ii) HTBP   (iii) OTSO   (iv) LFEL
(v) OKHP   (vi) HFSP.
(*b*) For the same lattice, what are the direction indices for the following directions?
(vii) OJ   (viii) LF   (ix) SM   (x) NT.

1.4 Using a cubic unit cell, obtain the angles between the normals to the pairs of planes defined by the following Miller indices:
(i) (100) (010)   (ii) (100) (111)   (iii) (100) (210)
(iv) (110) (210).

2 Introducing Tensors

2.1 Introduction to the Notation

Because a single crystal is inherently symmetrical, it presents the same aspect from a number of different directions. The properties of single cystals will generally depend on the direction in which the properties are measured. However, as a result of the crystal symmetry, there will be different directions in which the physical properties of the crystal are the same. Great simplification is obtained in any formulation of the physical properties of the crystal by exploiting the inherent symmetry.

In order to understand how to represent physical properties in crystals and to see why we need to develop a systematic approach, we will take as an example electrical resistivity. Assuming that we have an electric current density J passing through a crystalline sample when an electric field E is established within the sample, then these quantities are related by

$$E = \rho J$$

where ρ is the electrical resistivity of the sample. Provided the sample is isotropic (i.e. its properties are the same in all directions), the electric current and the electric field will be in the same direction and ρ is a scalar. However, this condition may not necessarily be so. E and J may not be parallel; in which case ρ cannot be represented by a scalar. Such a situation may occur when electric charge flows most easily in a direction which is different to that in which the electric field is applied. Resolving the electric field into components in different directions can then lead to components of current in the high-resistivity and low-resistivity directions. When combined, these will produce an overall current which is rotated compared with the applied electric field.†

Mathematically this is written

$$E_1 = \rho_{11}J_1 + \rho_{12}J_2 + \rho_{13}J_3$$
$$E_2 = \rho_{21}J_1 + \rho_{22}J_2 + \rho_{23}J_3 \qquad (2.1)$$
$$E_3 = \rho_{31}J_1 + \rho_{32}J_2 + \rho_{33}J_3$$

where we have assumed linearity between E components and J components, which is equivalent to assuming that Ohm's law holds within the sample. E_1, E_2 and E_3 are components of vector E for the electric field in rectangular Cartesian coordinates and similarly J_1, J_2 and J_3 are the components of J.

An alternative representation is in matrix form:

$$\begin{pmatrix} E_1 \\ E_2 \\ E_3 \end{pmatrix} = \begin{pmatrix} \rho_{11} & \rho_{12} & \rho_{13} \\ \rho_{21} & \rho_{22} & \rho_{23} \\ \rho_{31} & \rho_{32} & \rho_{33} \end{pmatrix} \begin{pmatrix} J_1 \\ J_2 \\ J_3 \end{pmatrix} \qquad (2.2)$$

and an even simpler representation is

$$E_p = \sum_{q=1}^{3} \rho_{pq} J_q = \rho_{pq} J_q. \qquad (2.3)$$

By convention, when the dummy suffix (the repeated letter q) appears in the right-hand form of equation (2.3), then summation of terms is implied with the dummy suffix taking values 1, 2 and 3. Hence it can be seen clearly that forms (2.1), (2.2) and (2.3) for the relationship between E and J are equivalent. In due course, we shall see that writing down resistivity in the form of nine components is expressing resistivity as a so-called second-rank tensor.

2.2 Reducing the Number of Components

Experience in the laboratory would suggest to us that many of

---

†A good, although not perfect, everyday analogy is the tea trolley. Assume that it is pushed in a forward direction but that all the wheels have been set at an angle displaced to this forward direction. Pushing the trolley will give it movement which will be somewhat angled to the forward direction because the trolley tends, even with some slipping, to travel preferentially more in the direction of the angled wheels.

the single crystals which we use do not have nine independent components for resistivity. Certain of the components will be zero, others will not be independent and may be equal to each other.

Consider the situation for a simple crystal structure such as the hexagonal close-packed structure exhibited by cadmium and illustrated in figure 2.1. By considering currents flowing in turn in different equivalent directions and using the set of equations (2.1), we can see how to reduce the number of components.

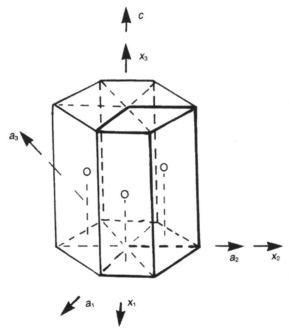

Figure 2.1 Axes and vector directions in a hexagonal crystal.

Consider the hexagonal crystal shown in figure 2.1 having lattice vectors a_1, a_2 and a_3 (all equal in magnitude to a) at 120° to each other in the basal plane and the c lattice vector perpendicular to the basal plane. The Cartesian coordinates are taken such that x_2 is parallel to a_2 and x_3 is parallel to c. We will consider currents of current density J flowing with resultant electric fields as follows:

Current, J	Resultant electric fields
Along a_1	$E_3 = \rho_{31}J_1 + \rho_{32}J_2$ $= \rho_{31}J\cos 30° + \rho_{32}J\cos 120°$ $= (\sqrt{3}\rho_{31}/2 - \rho_{32}/2)J$
Along a_2	$E_3 = \rho_{31}J_1 + \rho_{32}J_2 = \rho_{32}J$
Along a_3	$E_3 = \rho_{31}J_1 + \rho_{32}J_2$ $= -\rho_{31}J\cos 150° + \rho_{32}J\cos 180°$ $= (-\sqrt{3}\rho_{31}/2 - \rho_{32}/2)J$

All the above are equivalent hence

$$E_3 = 0 \text{ and } \rho_{31} = \rho_{32} = 0.$$

Along a_1	$E_{a_1} = E_1\cos 30° - E_2\cos 60°$ $= \rho_{12}J\cos 30° - \rho_{22}J\cos 60°$ $= (\sqrt{3}\rho_{12}/2 - \rho_{22}/2)J$
Along a_2	$E_{a_3} = E_1\cos 150° + E_2\cos 120°$ $= \rho_{12}J\cos 150° + \rho_{22}J\cos 120°$ $= (-\sqrt{3}\rho_{12}/2 - \rho_{22}/2)J$
Along a_3	$E_{a_1} = E_1\cos 30° + E_2\cos 120°$ $= \rho_{11}J\cos 150°\cos 30°$ $\quad + \rho_{22}J\cos^2 120°$ $= (-3\rho_{11}/4 + \rho_{22}/4)J$

All the above are equivalent: $\rho_{12} = -\rho_{12} = 0,\ -\rho_{22}/2 = -3\rho_{11}/4$ $+ \rho_{22}/4$, i.e. $\rho_{11} = \rho_{22}$.

By considering other equivalent configurations, we end up eventually with two independent components only:

$$\begin{bmatrix} \rho_{11} & 0 & 0 \\ 0 & \rho_{11} & 0 \\ 0 & 0 & \rho_{33} \end{bmatrix}.$$

Thus we have a resistivity component for any direction in the basal plane and another component for the c direction. This is as we might expect. However, the above is tedious and becomes more so with more complex crystals and properties. What we require is a formal way of obtaining the independent components and then using these component values to calculate the magnitudes of different physical effects in different crystalline directions.

2.3 Transformation of Axes

The orientation of the rectangular Cartesian axes we have been using can be chosen arbitrarily. So we can rotate our axes in any direction, but if we do this the physical properties of the crystal must stay the same as they cannot depend on the coordinate system. We can find the relationships between the components of the property before and after the rotation.

Let us take as the common origin of our axes a point O (figure 2.2) and have original axes Ox_1, Ox_2 and Ox_3 and new axes of Ox'_1, Ox'_2 and Ox'_3. Superscript $'$ is used to distinguish a new axis from an old (original) axis. In general terms, l_{ip} represents the cosine of the angle between the axis p of the old axes and the axis i of the new axes. We can now construct a table showing the so-called direction cosines. The direction cosine of x'_2 with respect to x_1 is l_{21} and so on:

	x_1	x_2	x_3
x'_1	l_{11}	l_{12}	l_{13}
x'_2	l_{21}	l_{22}	l_{23}
x'_3	l_{31}	l_{32}	l_{33}

Not all the ls are independent. We can realise this from the fact that the mutual orientation of the two sets of axes can be specified by three parameters, say three angles, which are often taken as the so-called Euler angles. Thus the dependence of the ls can be expressed as

$$l_{ik}l_{pk} = \delta_{ip}$$

where $\delta_{ip} = 1$ if $i = p$ and $\delta_{ip} = 0$ if $i \neq p$.

In the table above, if the cosines are calculated in a particular example, it is a good idea to check that the squares of the cosines (i.e. the l^2 terms) add to 1 across the rows and down the columns.

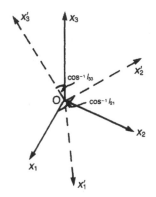

Figure 2.2 Rotation of axes.

2.4 Transformation of a Vector

A vector such as electric field E will have components E_p relative to axes Ox_p; that is, it will have E_1 relative to Ox_1, E_2 relative to Ox_2 and E_3 relative to Ox_3. Similarly, the vector will have components E_i' relative to Ox_i'. We can find the relationship between the vector related to one set of axes and the vector related to a second set of axes by thinking of E_1, E_2 and E_3 as vectors in the directions x_p (i.e. x_1, x_2 and x_3) and then resolving in the directions x_i' (where $i = 1$, 2, 3) to give E_i'. Thus we obtain

$$E_i' = E_1 \cos(x_1 \hat{O} x_i') + E_2 \cos(x_2 \hat{O} x_i') + E_3 \cos(x_3 \hat{O} x_i')$$
$$= l_{i1} E_1 + l_{i2} E_2 + l_{i3} E_3$$
$$= l_{ip} E_p. \tag{2.4}$$

In the form shown in the last line the neighbouring p suffixes are assumed to mean that we sum three terms (with $p = 1$, 2 and 3) in order to obtain expressions for any E_i'.

2.5 Transformation of the Coordinates of a Point

This is a simple extension of the transformation of a vector as the coordinates of a point can be considered as the coordinates of a vector originating from the origin and travelling out to that point. Thus we have coordinate x'_1, the coordinate for the point in the x'_1 axis direction, given by

$$x'_1 = l_{11}x_1 + l_{12}x_2 + l_{13}x_3.$$

(Compare with equation (2.1)).

We can write out similar expressions for x'_2 and x'_3 or we can write the whole out in a very much simpler short-hand notation in the form

$$x'_i = l_{ip}x_p. \tag{2.5a}$$

Again the neighbouring suffixes imply that for any x'_i we must add three terms with p taking values 1, 2 and 3.

Should we wish to transform back from new axes to old axes then we will have equations of the form

$$x_1 = l_{11}x'_1 + l_{21}x'_2 + l_{31}x'_3.$$

These can be abbreviated similarly to the form

$$x_p = l_{ip}x'_i. \tag{2.5b}$$

Note that we now have separated suffixes i, but summation of three terms with i taking values 1, 2 and 3 is still implied.

2.6 Transformation and Definition of a Tensor

Having considered how to transform a vector, we are able to extend the arguments to a physical property such as resistivity which we can classify as a second-rank tensor. First we consider the electric field E arising from current density J. Using E_p for electric field where p takes the values 1, 2 and 3 and J_q for current density where q also takes values 1, 2 and 3 for the three axes, we write resistivity in the form

$$E_p = \rho_{pq}J_q$$

as used previously in equation (2.3). For the electric field and the current density represented relative to a new set of axes we can write

$$E'_i = \rho'_{ij}J'_j.$$

We can relate the electric field components of the new (primed) set of axes to the old axes by the relation

$$E_i' = l_{ip}E_p$$

$$= l_{ip}\rho_{pq}J_q \text{ using equation (2.3).}$$

Also, current density J_q can be obtained relative to the new axes by

$$J_q = l_{jq}J_j'$$

so we have

$$E_i' = l_{ip}l_{jq}\rho_{pq}J_q$$

$$= \rho_{ij}'J_j'.$$

Hence,

$$\rho_{ij}' = l_{ip}l_{jq}\rho_{pq}. \tag{2.6}$$

This shows how resistivity measured by components related to one set of axes can be transformed into resistivity measured by components related to a different set of axes by using direction cosines. Remember that we have dummy suffixes, and summations are implied. A quantity which transforms like resistivity is called a second-rank tensor by definition. The fact that it transforms in this way is the basic property of a tensor.

So a second-rank tensor representing a physical property can be defined as a quantity which transforms according to

$$T_{ij}' = l_{ip}l_{jq}T_{pq}. \tag{2.7}$$

The tensor will have nine components. Similarly a vector can be defined as a quantity which has three components (i.e. a first-rank tensor) and transforms according to

$$p_i' = l_{ip}p_p. \tag{2.8}$$

Extending further, we can say that a third-rank tensor has 27 components and transforms according to

$$T_{ijk}' = l_{ip}l_{jq}l_{kr}T_{pqr} \tag{2.9}$$

and so on.

We will not be going beyond fourth-rank tensors in this book, where, of the 81 components, fortunately only some will be independent when representing the properties of real crystals.

2.7 Second-rank Symmetrical Tensors; the Representation Quadric

Second-rank tensors are said to be symmetrical if $T_{pq} = T_{qp}$. Most common second-rank tensors describing physical properties are symmetrical, although a commonly occurring exception is the thermoelectric tensor. Alternatively if $T_{pq} = -T_{qp}$, the tensor is said to be antisymmetrical or skew symmetrical.

Second-rank symmetrical tensors can be represented in a most useful mathematical way, and then as a consequence, in a graphical way. They can be illustrated by using the so-called representation quadric. For this, we start by considering an equation of the form

$$S_{pq}x_px_q = 1 \qquad (2.10)$$

where p and q take values 1, 2 and 3. Let us write out the equation fully

$$S_{11}x_1^2 + S_{12}x_1x_2 + S_{13}x_1x_3 + S_{21}x_2x_1 + S_{22}x_2^2$$
$$+ S_{23}x_2x_3 + S_{31}x_3x_1 + S_{32}x_3x_2 + S_{33}x_3^2 = 1. \qquad (2.11)$$

If we let $S_{pq} = S_{qp}$ and collect like terms, the equation simplifies to

$$S_{11}x_1^2 + S_{22}x_2^2 + S_{33}x_3^2 + 2S_{21}x_2x_1 + 2S_{23}x_2x_3 + 2S_{31}x_3x_1$$
$$= 1. \qquad (2.12)$$

This is the equation of a surface where the xs represent distances along three Cartesian axes. The equation can be transformed to different axes using the relations for transformation of axes. Let us change to a set of new axes x'. Then

$$x_p = l_{ip}x_i'$$

and

$$x_q = l_{jq}x_j'$$

so

$$S_{pq}l_{ip}l_{jq}x_i'x_j' = 1$$

or

$$S_{ij}'x_i'x_j' = 1$$

where

$$S'_{ij} = l_{ip}l_{jq}S_{pq}. \tag{2.13}$$

Comparing equation (2.13) for S with equation (2.7) for T earlier, we see that S_{pq} transforms like the components of a second-rank symmetrical tensor (i.e. a tensor where $T_{pq} = T_{qp}$). So any second-rank tensor can be represented by a representation quadric of the form illustrated by equation (2.11).

Quadrics described by such an equation have principal axes at right angles such that they can be put into the simpler form of

$$S_1 x_1^2 + S_2 x_2^2 + S_3 x_3^2 = 1. \tag{2.14}$$

That is, a suitable transformation of the axes removes the cross-product terms in x. By analogy, a tensor will also have principal axes to which it can be referred. Transformation of axes in this way is not necessarily straightforward and is discussed further in Appendix 2.

So

$$\text{tensor } T_{pq} = \begin{bmatrix} T_{11} & T_{12} & T_{13} \\ T_{21} & T_{22} & T_{22} \\ T_{31} & T_{32} & T_{33} \end{bmatrix}$$

becomes

$$\begin{bmatrix} T_{11} & 0 & 0 \\ 0 & T_{22} & 0 \\ 0 & 0 & T_{33} \end{bmatrix}$$

with respect to the principal axes and is quite often written as

$$\begin{bmatrix} T_1 & 0 & 0 \\ 0 & T_2 & 0 \\ 0 & 0 & T_3 \end{bmatrix}$$

where T_1, T_2 and T_3 are the principal coefficients of the tensor $[T_{pq}]$. Use of the single subscript will be particularly useful and make especial sense when we consider fourth-rank tensors and a matrix notation later (see p.58).

If T_1, T_2 and T_3 are all positive, then the tensor can be represented by an ellipsoid whose semiaxes have lengths $1/\sqrt{T_1}$, $1/\sqrt{T_2}$ and $1/\sqrt{T_3}$ (see figure 2.3(a)). If two of the principal components are positive and one is negative, the surface is a hyperboloid of one sheet (figure 2.3(b)), if one is positive and two are negative it is a hyperboloid of two sheets (figure 2.3(c)), and if all three coefficients are negative, the surface is an imaginary ellipse. When considering phenomena such as conductivity and resistivity, we always have positive coefficients and a real ellipsoid. An example where this is not so, however, is when representing strain by a second-rank tensor. In this case, we can have either tensile or compressive strains, and thus any of the listed possibilities.

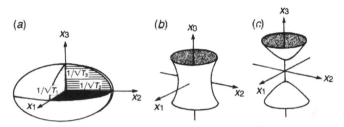

Figure 2.3 The representation quadric for a second-rank tensor T_{pq}.

It is necessary to relate the representation quadric and hence the second-rank symmetrical tensor to the situation in real crystals. Table 2.1 summarises this. It takes the crystal systems in order of decreasing overall symmetry. In Chapter 5 we shall be considering optical properties of crystals so the optical classification is included in the second column. The third column gives details of the representation quadric including, where necessary, the orientation of the quadric. Only three components are ever necessary to define a second-rank tensor when referred to its principal axes. However, if we need to specify the directions of these axes then another three coefficients are required. So in the application of tensors to monoclinic and triclinic crystals, coefficients other than the principal components are necessary in order to relate the tensors to the conventional crystallographic axes. It would be possible to rotate the axes of the tensor such that only the three principal coefficients are necessary but we would have no information regarding the orientation of the representation ellipsoid relative to the crystallographic axes. In the case of the representation

Table 2.1 Relationship between second-rank tensors and crystal systems.

Crystal system	Optical classification	Representation quadric	Number of independent coefficients	Tensor referred to conventional axes
Cubic	Anaxial	Sphere	1	$\begin{bmatrix} T & 0 & 0 \\ 0 & T & 0 \\ 0 & 0 & T \end{bmatrix}$
Tetragonal hexagonal trigonal†	Uniaxial	Quadric of revolution about $x_3 = z$	2	$\begin{bmatrix} T_1 & 0 & 0 \\ 0 & T_1 & 0 \\ 0 & 0 & T_3 \end{bmatrix}$
Orthorhombic	Biaxial	x_1, x_2, x_3 parallel to diads	3	$\begin{bmatrix} T_1 & 0 & 0 \\ 0 & T_2 & 0 \\ 0 & 0 & T_3 \end{bmatrix}$
Monoclinic	Biaxial	x_2 parallel to diad	4	$\begin{bmatrix} T_{11} & 0 & T_{13} \\ 0 & T_{22} & 0 \\ T_{13} & 0 & T_{33} \end{bmatrix}$
Triclinic	Biaxial	General quadric	6	$\begin{bmatrix} T_{11} & T_{12} & T_{13} \\ T_{12} & T_{22} & T_{23} \\ T_{13} & T_{23} & T_{33} \end{bmatrix}$

† A hexagonal cell is used.

quadric for the monoclinic system the x_2 axis of the quadric is taken as the axis about which the crystal exhibits twofold rotational symmetry: i.e. x_2 is taken as the diad (twofold rotational) axis of the crystal system. In crystallography it is often referred to as the y axis. This results in the need for the T_{13} component but not the T_{23} component. For the orthorhombic system, the quadric axes are the same as the diad axes. For the other crystal systems, the orientation of the ellipsoid is self-evident as in each case the quadric axes conform with the standard crystallographic axes. We shall return to the use of the ellipsoid to represent second-rank symmetrical tensors in Chapter 3.

2.8 Neumann's Principle

For the higher-rank tensors in particular, if all the components were to be independent when we applied the tensors to the properties of real crystals, then calculations would become alarmingly complex. However, application of tensors to physical properties requires us to take fully into account the crystal symmetries discussed in Chapter 1. In particular, we make use of Neumann's principle which states: *The symmetry elements of any physical property of a crystal* must *include the symmetry elements of the point group of the crystal*.

It is important to note that the physical properties may, and often do, include more symmetry than the point group. Physical properties characterised by second-rank tensors must be centrosymmetric whereas the crystal itself may not be so. We can see this is the case if we have

$$p_i = T_{ij}q_j.$$

Reversing the directions of p and q changes the signs of each of their components but leaves T_{ij} unchanged.

2.9 Some Examples of Tensors

Second rank. The common examples of second-rank tensors are

electrical conductivity and electrical resistivity and thermal conductivity and thermal resistivity, and these will be considered in
detail in Chapter 3. Other examples are permittivity, electric
and magnetic susceptibility, magnetic permeability and diffusion.
They relate two vectors, and the tensors in all these cases are
symmetrical. One can also have second-rank tensors relating a
scalar and another second-rank tensor. Thermal expansion is an
example of this; the thermal expansion coefficient relates the
strain to the change of temperature. The thermoelectric effect is
an example of a non-symmetrical second-rank tensor relating
two vectors. These vectors are chemical potential gradient and
temperature gradient, and the thermoelectric tensor is not a
particularly straightforward example of a tensor. The Peltier
effect is an example of a non-symmetrical second-rank tensor
relating a scalar and another second-rank tensor (but again not
an especially straightforward example of a tensor property).
However, dividing tensors into groups according to what they
relate is somewhat arbitrary. If electrical resistivity were to be
defined not from Ohm's law but by relating Joule heating to
current density components, i.e. by $\dot{Q} = \rho_{pq} J_p J_q$, where $\dot{Q}$ is
the Joule heating, then it would relate a scalar to a second-rank
tensor. However, the rank of ρ_{pq} or of any other tensor is
independent of the defining equation. A further type of second-
rank tensor is illustrated by optical activity. It is a pseudo or
axial tensor and will be discussed in Chapter 6.

Third rank. These tensors relate a vector to a second-rank
tensor. A typical example is the piezoelectric effect in which
electric polarisation arises from stress in a crystal. Another
example is the Hall tensor, which is a further example of an
axial tensor and will be considered in Chapter 6. Alternatively,
it can be shown to be equivalent to a non-symmetrical second-
rank tensor.

Fourth rank. Usually these will relate two second-rank tensors. The classic example is elastic compliance (or its inverse
elastic stiffness) relating stress to strain. This example will be
discussed in detail in Chapter 4.

There is a further complication when considering optical
effects. Light passing through the crystal consists of transverse
waves whose vibrations can be resolved into two perpendicular
components, both orthogonal to the direction of motion of the

wave. The interaction of both components with the crystal must then be considered. This will be discussed in Chapter 5.

It is useful to combine the various types of tensors with the 32 crystal classes and to summarise the overall number of maximum independent coefficients which can occur. Table A1 in Appendix 1 is such a summary.

2.10 Worked Example Showing Change of Axes for a Tensor

Example 2.A
The electrical conductivity tensor for a certain hypothetical crystal has the components

$$\begin{bmatrix} 146 & -18 \times \sqrt{3} & 0 \\ -18 \times \sqrt{3} & 182 & 0 \\ 0 & 0 & 72 \end{bmatrix}$$

(in units of $10^7 \, \Omega^{-1} \mathrm{m}^{-1}$). Take new axes rotated 60° about x_3 in a clockwise direction looking along $-x_3$ and write down a table of direction cosines relating the new and old axes. Check that the squares of the l_{ip} in each row and column sum to 1. Write down the components of the conductivity tensor referred to the new axes and note that the tensor is now referred to principal axes.

Answer
Figure 2.4 shows the relative orientation of the two sets of axes. We draw up the table of direction cosines;

	x_1	x_2	x_3	Σl^2
x_1'	$1/2$	$-\sqrt{3}/2$	0	1
x_2'	$\sqrt{3}/2$	$1/2$	0	1
x_3'	0	0	1	1
Σl^2	1	1	1	

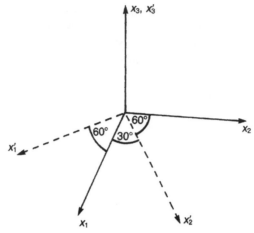

Figure 2.4 Orientation of axes for Question 2.A.

Next we use $\sigma'_{ij} = l_{ip}l_{jq}\sigma_{pq}$ to give

$$\sigma'_{11} = l_{11}^2\sigma_{11} + l_{11}l_{12}\sigma_{12} + l_{12}l_{11}\sigma_{21} + l_{12}^2\sigma_{22}$$

$$= (1/4)\,(146) + (1/2)(-\sqrt{3}/2)(-\sqrt{3}\times18)$$

$$+ (-\sqrt{3}/2)(1/2)(-\sqrt{3}\times18) + (3/4)(182)$$

$$= 200$$

$$\sigma'_{12} = l_{11}l_{21}\sigma_{11} + l_{11}l_{22}\sigma_{12} + l_{12}l_{21}\sigma_{21} + l_{12}l_{22}\sigma_{22}$$

$$= (\sqrt{3}/4)(146) + (\sqrt{3}/4)(18) + (3\sqrt{3}/4)(18)$$

$$- (\sqrt{3}/4)(182)$$

$$= 0$$

$$\sigma'_{21} = l_{21}l_{11}\sigma_{11} + l_{21}l_{12}\sigma_{12} + l_{22}l_{11}\sigma_{21} + l_{22}l_{12}s_{22}$$

$$= (\sqrt{3}/4)(146) + (3\sqrt{3}/4)(18) + (\sqrt{3}/4)(18)$$

$$- (\sqrt{3}/4)(182)$$

$$= 0$$

$$\sigma'_{22} = l_{21}^2\sigma_{11} + l_{21}l_{22}\sigma_{12} + l_{22}l_{12}\sigma_{21} + l_{22}^2\sigma_{22}$$

$$= (3/4)(146) - (\sqrt{3}/2)(\sqrt{3}/2)(18)$$

$$- (\sqrt{3}/2)(\sqrt{3}/2)(18) + (1/4)(182)$$

$$= 128.$$

The pair of terms σ'_{13} and σ'_{31} and the pair of terms σ'_{23} and σ'_{32} are zero because each term involves a direction cosine which is zero. σ'_{33} has the same value as σ_{33}.

The form of the tensor is

$$\begin{bmatrix} 200 & 0 & 0 \\ 0 & 128 & 0 \\ 0 & 0 & 72 \end{bmatrix}$$

There are diagonal components only; hence the tensor is referred to principal axes.

Problem

2.1 A tetragonal crystal has electrical resistivity components in units of $\mu\Omega$ m of

$$\begin{bmatrix} 16 & 0 & 0 \\ 0 & 10 & -6 \\ 0 & -6 & 10 \end{bmatrix}$$

referred to axes as follows: x_1 along an 'a' crystallographic axis; x_2 at 45° to the 'c' axis in the 'ac' plane; x_3 orthogonal to x_1 and x_2.

Find the electrical resistivity components referred to the principal axes.

3 Second-rank Tensors; Conductivity

3.1 Thermal Conductivity and Thermal Resistivity

In an isotropic medium, heat conduction can be represented by the equation

$$h = - k \text{ grad } T.$$

k is the thermal conductivity and is a constant, having the same value in whatever direction it is measured in the medium.

$$\text{grad } T = \partial T / \partial x_1 + \partial T / \partial x_2 + \partial T / \partial x_3$$

and is the temperature gradient. The heat flow h per unit area is in the direction of greatest fall of temperature, i.e. in the direction of $- \text{grad } T$. In suffix notation, we can write the equation for heat conduction as

$$h_i = - k \partial T / \partial x_i.$$

In a crystal the conductivity may take different magnitudes if measured in different directions, i.e. it may vary as a function of crystallographic direction. Hence the heat flow may not be in the same direction as the maximum temperature gradient. We now write the conductivity equation as

$$h_i = - k_{ij} \partial T / \partial x_j \tag{3.1}$$

where k_{ij} is the second-rank thermal conductivity tensor. Alternatively we can use the thermal resistivity tensor r_{ij} and write

$$\partial T / \partial x_i = - r_{ij} h_j. \tag{3.2}$$

It can be argued that k_{ij} and r_{ij} are symmetrical tensors but the argument for this is not as straightforward as it is for some second-rank tensors representing physical properties. One approach (see Nye 1957, p.204) is to take a flat disc whose outer circular surface is maintained isothermally. The disc is cut from a crystal such that the axis x_3 perpendicular to the plane of the disc has crystallographic rotational symmetry. It is concluded that if $k_{ij} \neq k_{ji}$, heat must *spiral* out from the centre of the disc even though the symmetry of the set-up ensures circular isotherms centred around axis x_3. No experimental evidence has been obtained for such spiralling of the heat flow. Hence, the conductivity tensor and the resistivity tensor are symmetrical.

If in addition we refer the tensors to principal axes, we obtain the tensors in the forms

$$\begin{bmatrix} k_1 & 0 & 0 \\ 0 & k_2 & 0 \\ 0 & 0 & k_3 \end{bmatrix} \quad \text{and} \quad \begin{bmatrix} r_1 & 0 & 0 \\ 0 & r_2 & 0 \\ 0 & 0 & r_3 \end{bmatrix}.$$

We can use the representation quadric for symmetrical second-rank tensors as discussed in Chapter 2 and use conductivity and resistivity ellipsoids to represent the thermal properties of our crystal.

3.2 Heat Flow in Crystal Samples

Let us consider heat flow in two samples with different geometrical shapes. Example (*a*) in figure 3.1 shows a very thin sample with large heat source and heat sink on opposite faces. It is equivalent to the classical Lees' disc experiment for measuring the thermal conductivity of an isotropic sample which is a poor thermal conductor. Since the sample is of large cross section compared with its thickness, the isothermal surfaces must be parallel to the end faces of the sample. The temperature gradient must be perpendicular to the end faces. If the sample is single crystal and not isotropic, then from what we said in §3.1, the heat flow will not necessarily be in this same direction.

Example (*b*) consists of a long narrow (often cylindrical) sample with heat source and sink at opposite ends. It is equivalent to the classical Searle's method for measuring the conductivity of a good conductor. The side walls are lagged, or alternatively the sample is arranged in an evacuated chamber to prevent heat loss from the side walls. In this case, it is the heat flow which is along the length of the sample, and the temperature gradient which may be at some oblique direction.

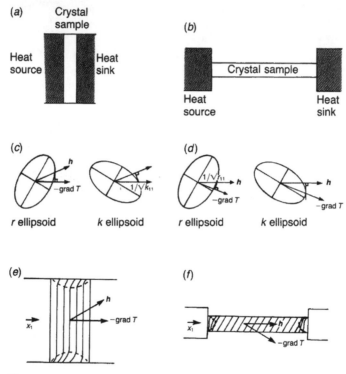

Figure 3.1 Heat flow in (*a*) a thin wide sample and (*b*) a long narrow sample. (*c*) and (*d*) show the cross sections of the corresponding conductivity and resistivity ellipsoids and (*e*) and (*f*) show the pattern of isothermals in the two cases.

Let us assume for the two cases that we have crystals which are mounted with similar orientation between the heater and sink. We will assume that one principal direction for the conductivity or resistivity lies perpendicular to the plane of the paper and that the other two principal directions lie at angles to the direct line between heater and sink. We can draw the corresponding cross sections of the conductivity and resistivity ellipsoids for the two cases (figures 3.1(c) and (d)).

Looking at 3.1(c) we can see that the orientation of the crystal is such that heat flows preferentially in a direction sloping upwards (i.e. up the page) relative to the temperature gradient. The conductivity ellipsoid is tilted in such a way that the thermal conductivity maximises at an angle measured in an anticlockwise direction from $-\text{grad } T$ as we look down onto the page. Note that as the heat flow is angled to the sides of the sample, the isothermal surfaces cannot remain perpendicular to the sides right to the edges. Hence figure 3.1(e) shows the pattern of isothermals which arises in the experimental configuration. The equation suitable for this set-up is the one which obtains heat flow as a consequence of established temperature gradient. Hence the conductivity equation (equation (3.1)) is applicable and the set-up is that to measure thermal conductivity in an anisotropic crystal. Remember that the semiaxes of the ellipsoids have magnitude $1/\sqrt{T_p}$ where T_p is a principal value of the tensor property.

By comparison, in the configuration shown in figure 3.1(d), the temperature gradient will be displaced in a clockwise direction relative to h as once again we look down onto the page. In this case, it is the isothermals which are angled relative to the sides of the sample. There will be distortion of the directions of the isothermals close to the heat source and sink as shown in 3.1(f). This set-up is the one required to measure thermal resistivity and equation (3.2) is the correct one to apply. Note that if principal values of conductivity are required these can be calculated once the principal values of resistivity are known, because of the reciprocal relationship between *principal* components.

If the samples used are oriented with their longitudinal axis along a principal direction, the reciprocal relationship can be used immediately and there is no difficulty as to which configuration (that of figure 3.1(a) or of figure 3.1(b)) should be used, as temperature gradient and heat flow are aligned. It is common to cut samples in this way to simplify interpretation of experimental results. However, this is not always experimentally possible. Growth of single crystals is often in preferred directions, which can make cutting to specific shapes more difficult. If samples of arbitrary orientation are used, the conductivity components can be obtained from resistivity components only when sufficient measurements have been made to determine each principal component of resistivity. This latter will require the use of a number of differently oriented samples unless the material is isotropic or cubic. Once the measurements have been made, it is necessary to use the tensor equations as outlined here in order to obtain values of the principal components. It is these values which are usually quoted when listing crystalline properties and which are the most useful when making further studies.

3.3 The Radius-normal Property of the Representation Quadric

This property can be stated as follows: if T_{ij} are the components of a symmetrical second-rank tensor relating the vectors p and q so that $p_i = T_{ij}q_j$, then the direction of p for a given q can be found by drawing a radius vector OQ of the representation quadric parallel to q and finding the normal to the quadric at Q.

This property is illustrated in figure 3.2. OQ is a line drawn from the centre of the cross section of the representation quadric, the ellipsoid, in the direction of q. It intersects the elliptical cross surface of the ellipsoid at Q. QN is the normal to the ellipsoid surface at Q and gives the direction of p, which can then be represented by a line OP parallel to QN. By geometry, the tangent to the ellipse at Q will meet the line OP at right angles.

We can prove the radius-normal property as follows.

(1) We take axes Ox_1, Ox_2 and Ox_3 as principal axes for our second-rank tensor T_{ij}, which then simplifies to the three principal components T_1, T_2 and T_3.

(2) Let the direction cosines of the vector q relative to the axes Ox_1, Ox_2 and Ox_3 be l_1, l_2 and l_3. If OQ has length r where Q lies on the ellipsoid, then the coordinates of Q are rl_1, rl_2 and rl_3. Q will satisfy the equation

$$T_1x_1^2 + T_2x_2^2 + T_3x_3^2 = 1 \qquad (3.3)$$

and the radius OQ will also have direction cosines l_1, l_2 and l_3.

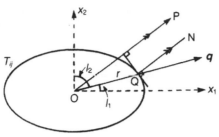

Figure 3.2 The radius-normal property of the representation quadric.

(3) It is a general property of an ellipsoid as defined by equation (3.3) that where the direction cosines of the radius are l_1, l_2 and l_3, the direction cosines of the normal are proportional to l_1T_1, l_2T_2 and l_3T_3. This is a standard result found in books of three-dimensional analytical geometry.

(4) Now q can be resolved into components along Ox_1, Ox_2 and Ox_3 proportional to l_1q, l_2q and l_3q, where q is the magnitude of q. This will give corresponding components for p proportional to T_1l_1q, T_2l_2q and T_3l_3q.

(5) Hence p has direction cosines proportional to l_1T_1, l_2T_2 and l_3T_3.

(6) Therefore p is parallel to the normal to the ellipsoid at Q.

3.4 Electrical Conductivity and Electrical Resistivity

The arguments regarding thermal conductivity and resistivity extend over to electrical conductivity and resistivity in an analogous manner. Whereas we had equation (3.1) for heat flow, we now have an equation for electric current (flow of charge) in an anisotropic medium given by

$$J_i = -\sigma_{ij}\partial V/\partial x_j = \sigma_{ij}E_j. \qquad (3.4)$$

J_i is the current density (i.e. the current per unit area of cross section) and should not be confused with the total current flowing. $\partial V/\partial x_j$ is the potential gradient through the sample; but it is more usual to use the form of the equation involving electric field components E_j, in which case the negative sign disappears due to the direction in which the electric field is defined. σ_{ij} is the electrical conductivity tensor. Equation (3.4) expresses Ohm's law and enables the calculation of current when an electric field is applied to the sample.

Alternatively, we can write

$$E_i = \rho_{ij} J_j \tag{3.5}$$

where ρ_{ij} is the resistivity tensor. The form is analogous to equation (3.2) for thermal resistivity, and enables us to calculate the electric field set up when an electric current flows.

Later in the book (Chapter 6) we shall consider the generalisation of the resistivity tensor in the presence of a magnetic field and consider the Hall effect tensor which arises.

3.5 Diffusion

This is a further example of a transport property which can be represented by a second-rank symmetrical tensor. In polycrystalline (isotropic) solids in which a concentration gradient of individual species of atoms occurs, then a flow of these atoms will take place; the magnitude of the flow will be proportional to the concentration gradient $\partial c/\partial r$. The flux of atoms will be given by

$$J = -D\partial c/\partial r \tag{3.6}$$

where D is the diffusion coefficient. It is usually found experimentally that its value is independent of the concentration gradient. Equation (3.6) is a statement of Fick's first law. The diffusion of the atoms is in the same direction as the concentration gradient (but of course moving towards the lower concentration).

When we come to consider single crystals, the atoms of the atomic species may no longer diffuse parallel to the concentration gradient present in the crystal. So now we need to represent the flow of atoms using an equation of the form

$$J_i = - D_{ij} \partial c / \partial x_j. \tag{3.7}$$

Measurement of the diffusion coefficients for atoms in crystals (usually the self-diffusion of atoms of one of the composite elements of the material) can give important information regarding crystalline structure.

3.6 Some Worked Examples of Second-rank Tensor Properties

Question 3.A Thermal conductivity and thermal resistivity of samples

Samples are set up as shown in figures 3.1(a) and 3.1(b). They are cut from a crystal of tetragonal structure with $c = 1.4a$. The heat source and heat sink for each sample are attached to {101} faces. The crystal has principal conductivities $k_{11} = 15 \ \text{W m}^{-1}\text{K}^{-1}$ and $k_{33} = 10 \ \text{W m}^{-1}\text{K}^{-1}$. In ($a$) the sample is 0.4 cm thick and faces are circular of area 10 cm². If the temperature gradient is $50 \, °\text{C m}^{-1}$, what is the heat flow between source and sink? In (b) the sample is 10 cm long and has a cross sectional area of 1 cm². Heat flow is 1 W. What is the temperature difference established between heat source and heat sink?

Answer
Sample (a): Sample orientation is shown in figure 3.3. The direction of the arrow in this case indicates the maximum temperature gradient, dT/dx. (Note that sample thickness of 0.4 cm is quoted to justify this, and that thickness is not required in the calculation.)

We require the value of the thermal conductivity k_{xx} for the direction x.

$$k_{xx} = l_{x1}^2 k_{11} + l_{x3}^2 k_{33}$$

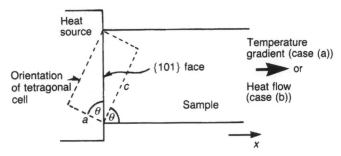

Figure 3.3 Sample orientation for Question 3.A.

where k_{11} refers to the 'a' direction of the tetragonal crystal and k_{33} refers to the 'c' direction. l_{x3} ($= \sin \theta$) is the cosine of the angle between the x direction and the principal axis 1 (corresponding to 'a' of the tetragonal cell) and l_{x3} ($= \cos \theta$) is the cosine of the angle between the x direction and the principal axis x_3 (corresponding to 'c' of the tetragonal cell).

$$\tan \theta = c/a = 1.4 \qquad \theta = 54.5°$$

$$\cos^2 \theta = 0.338 \qquad \sin^2 \theta = 0.662$$

$$k_{xx} = 0.662 \times 15 + 0.338 \times 10 = 13.3 \, \text{W} \, \text{m}^{-1} \text{K}^{-1}$$

We can now use

$$h_x = - k_{xx} \, \partial T / \partial x$$

$$= 13.3 \times 50 = 66.5 \, \text{W} \, \text{m}^{-2}$$

Total heat flow $= h_x \times$ cross sectional area

$$= 66.5 \, \text{mW}.$$

Sample (b): The direction of the arrow in figure 3.3 now refers to heat flow. We need to use $\partial T / \partial x = - r_{xx} h_x$ along the sample.

$$r_{11} = 1/k_{11} = 1/15$$

$$r_{33} = 1/k_{33} = 1/10 \, \text{(principal axes)}$$

$$r_{xx} = l_{x1}^2 r_{11} + l_{x3}^2 r_{33}$$

$$= \sin^2 \theta / 15 + \cos^2 \theta / 10 = 0.78 \, \text{W}^{-1} \text{m} \, \text{K}$$

Temperature difference $= \Delta T$

$$= 0.078 \times 10^4 \times 0.1 \,°C = 78 \,°C.$$

Question 3.B Electrical resistivity; angle between electric field and electric current

A long sample of a single crystal of uniform cross section is cut such that its length is in the [101] direction. The crystal has tetragonal structure with $c = 2a$ and has principal resistivities of $\rho_{11} = 16 \,\Omega \,m$ and $\rho_{33} = 4 \,\Omega \,m$. What is the resistivity along the length of the sample? Obtain the angle which the electric field makes with the current when an electrical current passes along the sample between the end faces.

Answer

We refer to figure 3.4(a). Let ρ_{xx} be the resistivity of the sample along its length.

Method 1 (calculation)

$$\rho_{xx} = l_{x1}^2 \rho_{11} + l_{x3}^2 \rho_{33}$$

$$l_{x1} = \sin \phi$$

$$l_{x3} = \cos \phi$$

$$\tan \phi = a/c = 0.5 \qquad \phi = 26.6\,°$$

$$\cos^2 \phi = 0.8 \qquad \sin^2 \phi = 0.2$$

$$\rho_{xx} = 16/5 + 4 \times 4/5 = 6.4 \,\Omega \,m$$

$$= \text{resistivity along the sample.}$$

Component of current along the 'a' direction

$$= I_a = 6.4 \sin \phi \times I = (6.4/\sqrt{5})I$$

where I is the total current along the sample.

Component of current along the 'c' direction

$$= I_c = 6.4 \sin \phi \times I = (6.4 \times 2/\sqrt{5})I.$$

Electric field along the 'a' direction

$$= E_a = \rho_{aa} I_a = (16 \times 6.4/\sqrt{5})I.$$

Electric field along the 'c' direction

$$= E_c = \rho_{cc}I_c = (4 \times 6.4 \times 2/\sqrt{5})I.$$

Angle between electric field E and 'a' axis is

$$\tan^{-1}(E_c/E_a) = \tan^{-1}0.5 = 26.6°.$$

Angle between I and 'a' axis is 63.4°.
Angle between E and $I = 36.8°$.

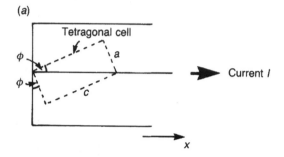

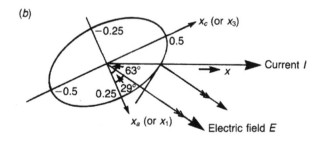

Figure 3.4 (*a*) Sample orientation for Question 3.B. (*b*) Resistivity ellipsoid for sample in Question 3.B.

Method 2 (graphical)
The representation quadric in the 'ac' plane is given by

$$S_a x_a^2 + S_c x_c^2 = 1.$$

Substituting values for the resistivity we obtain

$$16x_a^2 + 4x_c^2 = 1$$

x_a	x_c
0.0	±0.50
±0.10	±0.46
±0.20	±0.30
±0.23	±0.20
±0.25	0.0

Figure 3.4(b) shows the cross section in the 'ac' plane of the resistivity ellipsoid for the sample. Values for resistivity along the length of the sample and for the angle between I and E can be read off from the figure and agree with the calculation above. (For resistivity, measure the length of the radius vector of the ellipsoid in the direction of the field and obtain the square of the reciprocal of this value.)

Question 3.C Resistivity in an orthorhombic crystal

A current of 3 A flows between the ends of a single-crystal sample of gallium of length 2.5 cm. The crystal has square cross section of width 0.2 cm. Gallium has orthorhombic structure. Crystallographic study of this particular sample shows that it is oriented with its longitudinal axis at 35° to the 'c' crystallographic axis and at 57° to the 'b' crystallographic axis. The resistivity components for the gallium in units of $10^{-8}\,\Omega\,m$ are: along a axis, 55.5; along b axis, 17.3; along c axis, 7.85.

Calculate the voltage difference between the end faces of the sample.

Answer

If the longitudinal axis of the sample makes angles α, β and γ with the set of orthogonal axes defining the crystallographic axes a, b and c, then

$$\cos^2 \alpha + \cos^2 \beta + \cos^2 \gamma = 1$$

$$\cos^2 \alpha = 1 - \cos^2 57° - \cos^2 35°$$
$$= 1 - 0.297 - 0.671$$
$$= 0.0324$$

$$\cos \alpha = 0.180 \qquad \alpha = 79.6°.$$

We apply the general equation $E_i = \rho_{ij} J_j$ to the sample with both subscripts i and j referring to the longitudinal axis which we will label L and we calculate ρ_{LL}, the resistivity along the length of the crystal:

$$\rho_{LL} = (\cos^2 \alpha)\rho_a + (\cos^2 \beta)\rho_b + (\cos^2 \gamma)\rho_c$$

where ρ_a, ρ_b and ρ_c are the principal resistivities.

$$\rho_{LL} = (0.032 \times 55.5 + 0.297 \times 17.3 + 0.671 \times 0.785)10^{-8} \, \Omega \, m$$

$$= (1.78 + 5.14 + 5.27)10^{-8} \, \Omega \, m$$

$$= 1.2 \times 10^{-7} \, \Omega \, m.$$

Potential difference $= \rho_{LL} I L / A$

$$= (12.2 \times 10^{-8} \times 3 \times 2.5 \times 10^{-2})/(0.2 \times 0.2 \times 10^{-4}) \, V$$

$$= 2.29 \, mV.$$

Problems

3.1 A single-crystal quartz layer is used to thermally insulate a heated sample from a heat sink. The normal to the quartz layer lies at 45° to the 'a' and 'c' axes. The layer has square cross section with sides of length 2 cm and has thickness 0.3 cm. A uniform temperature difference of 75 °C is maintained between the large faces of the quartz layer. Calculate the rate of loss of heat from the sample through the quartz. The thermal conductivity of quartz is $6.5 \, W \, m^{-1} \, K^{-1}$ normal to the 'c' axis and $11 \, W \, m^{-1} \, K^{-1}$ along the 'c' axis.

3.2 A hexagonal cell can be used to describe the crystal structure of bismuth. The electrical resistivity of bismuth is $1.09 \times 10^{-6} \, \Omega \, m$ in the 'a' direction and $1.38 \times 10^{-6} \, \Omega \, m$ in the 'c' direction. A sample of bismuth is cut such that its longitudinal axes makes an angle of 26° with the 'c' axis. Calculate the resistivity along the axis of the sample. If a current passes longitudinally along the sample, calculate the angle between the direction of the current and the direction of the electric field within the sample (distant from the ends such that there are no end effects).

3.3 Gallium has orthorhombic crystal structure with $a \simeq b = 0.451$ nm and $c = 0.777$ nm. Using the resistivity data given in Example 3.C, calculate the electrical conductivity in the [111] direction.

3.4 Selenium has hexagonal crystallographic structure. The thermal conductivity normal to the 'c' axis is $1.4\,\mathrm{W\,m^{-1}\,K^{-1}}$ and parallel to the 'c' axis is $4.8\,\mathrm{W\,m^{-1}\,K^{-1}}$. Plot a cross section in the 'ac' plane for the conductivity ellipsoid. A thin crystal (thickness 4 mm) is cut with large $(01\bar{1}1)$ faces. A temperature difference of 30 °C is maintained between the faces. Using your plot, estimate both the rate of heat flow per cm^2 in a direction orthogonal to the end faces and the angle between the maximum rate of heat flow and the maximum temperature gradient. Note $a = 0.4355$ nm and $c = 0.4949$ nm.

4 Fourth-rank Tensors: Elasticity

4.1 Introduction

We will jump from second-rank tensors to fourth-rank tensors. We make this leap partly because properties involving third-rank tensors tend to arise less often, but mainly because the fourth-rank tensor for elasticity enables us conveniently to consider the techniques available for handling more complicated tensors. Elasticity is a physical property which is relatively easy to comprehend despite the large number of coefficients which can be involved. We can understand that if we apply forces (compressive or tensile) to a crystalline solid, there will be resulting deformation and that this deformation will not only depend on the directions of the forces but also on the ease with which the crystal deforms in particular directions due to underlying directional variations of structure. The physical property of elasticity, being represented by a fourth-rank tensor, relates two second-rank tensors, namely stress (i.e. a series of applied forces) and strain (the resulting deformation). Therefore we must first introduce these two second-rank tensors.

4.2 Strain

Strain is defined as the change in dimensions of a body as a result of subjecting the body to a system of forces which are in equilibrium. Strain may consist of an extension or a compression or alternatively it may be a shear.

For simplicity, we first consider linear extension in a thin wire. Let the wire have a length x between points O and P and a further small length Δx between P and Q (figure 4.1). We can then see what happens to the wire when we add a force to the far end of the wire keeping the other end (origin) fixed. Length x between O and P becomes $x + u$ where u is a small extension and measures the movement of P in a direction away from the origin O. The incremental length Δx between P and Q now becomes $\Delta x + \Delta u$ where Δu will be very small. Provided stretching of the wire is homogeneous, the extension u at any point P distance x along the wire will increase linearly with x. The strain in the particular length PQ is given by

$$\frac{\text{increase in length}}{\text{original length}} = \frac{\Delta u}{\Delta x}.$$

More particularly, the strain at P is given by

$$(\Delta u / \Delta x)_{x \to 0} = \mathrm{d}u / \mathrm{d}x = e \qquad (4.1)$$

where e is extension per unit length (relative change of length).

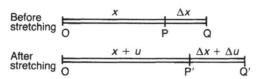

Figure 4.1 Extension of a thin wire.

Extending to a two-dimensional body, i.e. a plate, the situation immediately increases in complexity and we require four components. When we apply a system of forces to the plate, it will distort, and point P moves to P' and the initial infinitesimal lengths PQ_1 and PQ_2 before deformation become the infinitesimal lengths $P'Q'_1$ and $P'Q'_2$ (figure 4.2). Lengths PQ_1 and PQ_2 were taken orthogonal to each other and with their directions parallel to the x_1 and x_2 axes respectively. After the deformation, this will no longer be the case as can be seen in the figure. Whereas PQ_1 originally had length Δx_1 in the x_1 direction, after the deformation it has a resolved component of length $\Delta x_1 + \Delta u_1$ in the x_1 direction and a resolved component of length Δu_2 in the x_2 direction. It has rotated clockwise an angle θ relative to its original orientation. Depending on the forces deforming the plate, movement of PQ_2 will be somewhat different and independent of the movement of PQ_1, as shown in the figure.

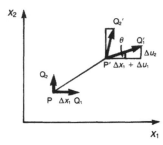

Figure 4.2 Distortion within a plate when subjected to a system of forces.

The extension per unit length of $P'Q_1'$ parallel to the x_1 direction is given by

$$e_{11} = \partial u_1/\partial x_1.$$

Similarly the extension per unit length parallel to the x_2 direction is given by

$$e_{22} = \partial u_2/\partial x_2.$$

We can also write down further terms of the form

$$e_{12} = \partial u_1/\partial x_2 \quad \text{and} \quad e_{21} = \partial u_2/\partial x_1.$$

These terms represent rotations. Referring to figure 4.2, we can see that

$$\tan \theta = \Delta u_2/(\Delta x_1 + \Delta u_1)$$
$$\simeq \Delta u_2/\Delta x_1$$

and so if displacements are small

$$\tan \theta = e_{21}.$$

e_{12} can also be expressed as an angle. But let us consider a pure rotation of our plate as a rigid sheet. The relative orientation of PQ_1 and PQ_2 remains 90° but PQ_1 and PQ_2 have rotated as a whole by angle ϕ (figure 4.3). e_{11} and e_{22} must be zero as there are no extensions. However, e_{21} is ϕ and e_{12} is $-\phi$ so that we have

$$[e_{ij}] = \begin{bmatrix} 0 & -\phi \\ \phi & 0 \end{bmatrix}. \tag{4.2}$$

This represents the pure rotation of the sheet and these components should not be part of the strain tensor. What we do is split our tensor $[e_{ij}]$ up into symmetrical and antisymmetrical parts.

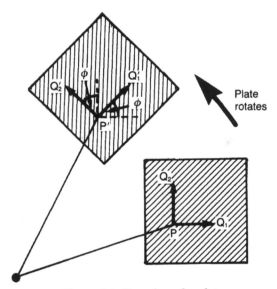

Figure 4.3 Rotation of a plate.

4.3 Symmetrical and Antisymmetrical Tensors

Any tensor can be divided up into the sum of two tensors, one symmetrical and the other antisymmetrical. Any single component of the original tensor can be put in the form

$$e_{ij} = \quad (e_{ij} + e_{ji})/2 \quad + \quad (e_{ij} - e_{ji})/2. \tag{4.3}$$

$$\text{symmetrical} \qquad \text{antisymmetrical}$$
$$\text{part} \qquad\qquad \text{part}$$

Thus tensor $[e_{ij}]$ is given by

$$\begin{bmatrix} e_{11} & e_{12} & e_{13} \\ e_{21} & e_{22} & e_{23} \\ e_{31} & e_{32} & e_{33} \end{bmatrix}$$

$$= \begin{bmatrix} e_{11} & (e_{12} + e_{21})/2 & (e_{13} + e_{31})/2 \\ (e_{12} + e_{21})/2 & e_{22} & (e_{23} + e_{32})/2 \\ (e_{13} + e_{31})/2 & (e_{23} + e_{32})/2 & e_{33} \end{bmatrix}$$

$$+ \begin{bmatrix} 0 & (e_{12} - e_{21})/2 & (e_{13} - e_{31})/2 \\ (e_{21} - e_{12})/2 & 0 & (e_{23} - e_{32})/2 \\ (e_{31} - e_{13})/2 & (e_{32} - e_{23})/2 & 0 \end{bmatrix}. \quad (4.4)$$

Sometimes it is useful to rotate the reference axes so as to make them coincide with the principal axes of the symmetrical part such that there are only the three diagonal terms in the symmetrical part. Hence e_{12} must equal $-e_{21}$ and so on. Doing this we get

$$\begin{bmatrix} e_{11} & 0 & 0 \\ 0 & e_{22} & 0 \\ 0 & 0 & e_{33} \end{bmatrix} + \begin{bmatrix} 0 & e_{12} & -e_{31} \\ -e_{12} & 0 & e_{23} \\ e_{31} & -e_{23} & 0 \end{bmatrix}$$

symmetrical part   antisymmetrical
referred to   part
principal axes

$$= \begin{bmatrix} e_{11} & e_{12} & e_{13} \\ -e_{12} & e_{22} & e_{23} \\ -e_{13} & -e_{23} & e_{33} \end{bmatrix}$$

$$= \begin{bmatrix} e_{11} & e_{12} & e_{13} \\ e_{21} & e_{22} & e_{23} \\ e_{31} & e_{32} & e_{33} \end{bmatrix}. \quad (4.5)$$

We have already referred to Appendix 2 for a discussion of diagonalisation.

4.4 The Strain Tensor

We can compare equations (4.2) and (4.3) and see that the rotation in (4.2) corresponds to the antisymmetrical part in (4.3). If we take out the antisymmetrical part and keep the symmetrical part, we eliminate the pure rotation and retain the strain only. We write the symmetrical part representing the strain in a plate in the form

$$[\varepsilon_{ij}] = \begin{bmatrix} \varepsilon_{11} & \varepsilon_{12} \\ \varepsilon_{21} & \varepsilon_{22} \end{bmatrix} = \begin{bmatrix} e_{11} & (e_{12} + e_{21})/2 \\ (e_{12} + e_{21})/2 & e_{22} \end{bmatrix}.$$

We can look at ε_{12} and ε_{21} and see that we have no rotation:

$$\varepsilon_{12} = (\partial u_1/\partial x_2 + \partial u_2/\partial x_1)/2$$

$$= \varepsilon_{21}$$

$$= 0.$$

We can go a step further if we wish and diagonalise the symmetrical part of the tensor.

In an analogous way, we can go on to define the strain tensor $[\varepsilon_{ij}]$ for a three-dimensional body as the symmetrical part of $[e_{ij}]$ in three dimensions where i and $j = 1, 2, 3$.

$$\begin{aligned} [\varepsilon_{ij}] &= \begin{bmatrix} \varepsilon_{11} & \varepsilon_{12} & \varepsilon_{13} \\ \varepsilon_{21} & \varepsilon_{22} & \varepsilon_{23} \\ \varepsilon_{31} & \varepsilon_{32} & \varepsilon_{33} \end{bmatrix} \\ &= \begin{bmatrix} e_{11} & (e_{12} + e_{21})/2 & (e_{13} + e_{31})/2 \\ (e_{12} + e_{21})/2 & e_{22} & (e_{23} + e_{32})/2 \\ (e_{13} + e_{31})/2 & (e_{23} + e_{32})/2 & e_{33} \end{bmatrix}. \end{aligned} \quad (4.6)$$

The three diagonal components are the tensile strains. If they have negative magnitude they become compressive strains. The other off-diagonal components measure the shear strains. As the principal strains can be either positive or negative, one cannot always represent strain by a real ellipsoid, something we could always do for resistivity and conductivity.

Strain is the response of the material, more particularly for our interests a crystal, to an influence. If this influence arises from some change within the crystal, such as a thermal expansion due to a uniform change of temperature in the crystal, then the influence itself will have no orientation. Hence the resulting strain must conform to the crystal symmetry. The strain will have directional properties because the thermal expansion coefficient will be different for different directions but Neumann's principle (p.31) must be obeyed. (Thermal expansion which will be considered briefly in Chapter 7 is a second-rank tensor.) In general, however, the strain will arise from external forces which can be applied with arbitrary magnitudes and directions. Whereas the resulting strain will be affected by the underlying crystal structure, the variation in strain will not arise purely from the underlying crystal structure. The strain tensor, and also the stress tensor, are examples of *field tensors* which are so-called because they do not represent crystal properties. Strictly, they are not encompassed by the title of this book. The resistivity and conductivity tensors, which we considered previously, do represent crystal properties and, as we have seen, conform to Neumann's principle and to the crystal properties. They are so-called *matter tensors*.

4.5 Stress

Any body which is acted on externally by a system of forces is in a state of stress. To consider the stress (i.e. the force per unit area) within a body it is usual to consider an element of volume, and for convenience we will consider a cube of material within our body. Provided the stress is homogeneous, the forces acting on the surfaces of this element will be independent of the position of this element within the body. We shall assume this. The assumption is not, of course, the same as assuming that the stresses are isotropic, and in fact anisotropic forces will be allowed for.

Figure 4.4(a) shows a cubic element within our body. The normal and shear components of the stresses are shown for three of the six faces. σ_{11}, σ_{22} and σ_{33} are the normal components and σ_{12}, σ_{23}, etc, are the shear components. The sign convention used is that if σ_{11}, etc, are positive, the stress is tensile, whereas if σ_{11}, etc, are negative, the stress is compressive. We assume that our body is in dynamic equilibrium, i.e. it is stationary with no rotation. For this to be so the forces on opposite faces must be equal and opposite. Also, if we look at figure 4.4(b), which takes a cross section in the x_2x_3 plane, we can see that for no turning moment $\sigma_{32} = \sigma_{23}$ and in general $\sigma_{ij} = \sigma_{ji}$. It is possible to prove that σ_{ij} is a second-rank tensor (see for instance Nye 1957, p.87) relating resultant stress with direction. We shall assume this.

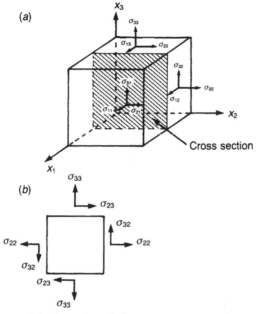

Figure 4.4 (*a*) Normal and shear components of stress on a cubic element. (*b*) Cross section of the cube in the $x_2 x_3$ plane. For no turning moment $\sigma_{ij} = \sigma_{ji}$.

4.6 Elasticity

We start with the well known law that for a solid body the strain is proportional to the magnitude of the stress, provided that the magnitude of the stress is below that corresponding to the elastic limit of the material such that the strain is recoverable. This is Hooke's law, usually expressed for an isotropic medium in the form

$$\sigma = c\varepsilon$$

where c is Young's modulus. We shall refer to c as elastic stiffness. Alternatively, we can write

$$\varepsilon = s\sigma$$

where s is elastic compliance. It is unfortunate that the convention used for stiffness and compliance involves symbols c and s which are the reverse of their initial letters.

We have already introduced stress σ and strain ε as second-rank tensors so we immediately write Hooke's law in the more general forms

$$\sigma_{ij} = c_{ijkl}\varepsilon_{kl} \qquad (4.7)$$

and

$$\varepsilon_{ij} = s_{ijkl}\sigma_{kl} \qquad (4.8)$$

where c_{ijkl} is the fourth-rank tensor for stiffness and s_{ijkl} is the fourth-rank tensor for compliance. That they are fourth-rank tensors can be demonstrated by showing that the 81 components for each transform on change of axes according to the transformation rule for a fourth-rank tensor. The rule is merely an extension of that for a third-rank tensor and therefore is an extension of equation (2.9). Although in general we would expect 81 components, we already have that

$$\sigma_{kl} = \sigma_{lk} \text{ for no turning moment on the crystal}$$

and

$$\varepsilon_{ij} = \varepsilon_{ji} \text{ from removal of pure rotation}$$

so

$$c_{ijkl} = c_{jikl} = c_{ijlk} = c_{jilk}$$

and

$$s_{ijkl} = s_{jikl} = s_{ijlk} = s_{jilk}.$$

Thus the number of independent components is reduced to 36 and it is possible to use a 6×6 matrix notation.

4.7 The Matrix Notation

This approach is possible because of the symmetry of c_{ijkl} and s_{ijkl} in their first and last pairs of suffixes. These pairs of suffixes are numbered singly from 1 to 6 by the following convention:

Tensor notation suffixes	11	22	33	23	32	31	13	12	21
Matrix notation suffixes	1	2	3	4		5		6	

As a consequence we shall wish to express the relationships between stress σ and strain ε in the matrix forms

$$\sigma_m = c_{mn}\varepsilon_n \qquad (4.9a)$$

and

$$\varepsilon_m = s_{mn}\sigma_n \qquad (4.9b)$$

where $m, n = 1, 2, \ldots 6$.

Thus, the principal components of stress σ become σ_1, σ_2 and σ_3, and we have similar subscript notation for stress. This fits with the single-subscript notation we referred to in Chapter 2 (§2.7) when, in considering conductivity and resistivity in high-symmetry crystals, we referred to tensor components corresponding to their principal axes.

There is a serious complication. If we redefine the pairs of suffixes exactly as indicated so far, we do not end up with equations in precisely the form of (4.9a) and (4.9b). Rather we should get factors of 2 and 4 occasionally included. This comes about because there will be more pairs of type 23 than 11, for instance. The conventional way around this is to transform stress to its matrix form in a direct way

$$\begin{bmatrix} \sigma_{11} & \sigma_{12} & \sigma_{13} \\ \sigma_{21} & \sigma_{22} & \sigma_{23} \\ \sigma_{31} & \sigma_{32} & \sigma_{33} \end{bmatrix} \rightarrow \begin{pmatrix} \sigma_1 & \sigma_6 & \sigma_5 \\ \sigma_6 & \sigma_2 & \sigma_4 \\ \sigma_5 & \sigma_4 & \sigma_3 \end{pmatrix}$$

but to transform strain to its matrix form according to

$$\begin{bmatrix} \varepsilon_{11} & \varepsilon_{12} & \varepsilon_{13} \\ \varepsilon_{21} & \varepsilon_{22} & \varepsilon_{23} \\ \varepsilon_{31} & \varepsilon_{32} & \varepsilon_{33} \end{bmatrix} \rightarrow \begin{pmatrix} \varepsilon_1 & \varepsilon_6/2 & \varepsilon_5/2 \\ \varepsilon_6/2 & \varepsilon_2 & \varepsilon_4/2 \\ \varepsilon_5/2 & \varepsilon_4/2 & \varepsilon_3 \end{pmatrix}.$$

In addition we need to introduce factors of 2 and 4 into the equations relating compliance in tensor and martrix notations

$$s_{ijkl} = s_{mn} \qquad \text{for } m = 1, 2, 3 \text{ and } n = 1, 2, 3$$

$$2s_{ijkl} = s_{mn} \qquad \text{for } either \ m \text{ or } n = 4, 5, 6$$

$$4s_{ijkl} = s_{mn} \qquad \text{for } both \ m \text{ and } n = 4, 5, 6.$$

However,

$$c_{ijkl} = c_{mn} \qquad \text{for all } i, j, k, l \text{ and all } m, n$$

(i.e. for $i, j, k, l = 1, 2, 3$ and $m, n = 1, \ldots 6$).

Table 4.1 demonstrates that the above equalities between tensor and matrix notations work out in practice. Other conventions are possible (sometimes the factors are incorporated within s_{mn}) and the reader should check when referring to other sources which convention is in use. However, the above approach is common.

We can now write out matrix arrays for c_{mn} and s_{mn} in square matrices;

$$\begin{vmatrix} c_{11} & c_{12} & c_{13} & c_{14} & c_{15} & c_{16} \\ c_{21} & c_{22} & c_{23} & c_{24} & c_{25} & c_{26} \\ c_{31} & c_{32} & c_{33} & c_{34} & c_{35} & c_{36} \\ c_{41} & c_{42} & c_{43} & c_{44} & c_{45} & c_{46} \\ c_{51} & c_{52} & c_{53} & c_{54} & c_{55} & c_{56} \\ c_{61} & c_{62} & c_{63} & c_{64} & c_{65} & c_{66} \end{vmatrix}$$

and

$$\begin{vmatrix} s_{11} & s_{12} & s_{13} & s_{14} & s_{15} & s_{16} \\ s_{21} & s_{22} & s_{23} & s_{24} & s_{25} & s_{26} \\ s_{31} & s_{32} & s_{33} & s_{34} & s_{35} & s_{36} \\ s_{41} & s_{42} & s_{43} & s_{44} & s_{45} & s_{46} \\ s_{51} & s_{52} & s_{53} & s_{54} & s_{55} & s_{56} \\ s_{61} & s_{62} & s_{63} & s_{64} & s_{65} & s_{66} \end{vmatrix}.$$

It is very important to remember that c_{mn} and s_{mn} are not now tensors and *do not transform as tensors*.

The maximum number of independent components is reduced further as the stiffness and compliance matrices are symmetrical, i.e. c_{mn} is the same as c_{nm}, and s_{mn} is the same as s_{nm}. This can be demonstrated by energy considerations. For instance we can show that s_{2211} is the same as s_{1122}. If we apply two stresses σ_1 and σ_2 to a volume element (a unit cube) of our crystal, the final strain energy should be the same irrespective of the order in which we apply the stresses. Applying a stress σ_m to the cube with consequential change of strain $d\varepsilon_m$ leads to a change of strain energy of

$$dW = \sigma_m d\varepsilon_m$$

$$= \sigma_m s_{mn} d\sigma_n$$

where m takes values 1, 2, ... 6. In the following examples we

Table 4.1 Examples of the relationships between matrix and tensor forms of the stress and strain components.

(i) Stress $\sigma_{ij} = c_{ijkl}\varepsilon_{kl}$

$\sigma_{11} = c_{1111}\varepsilon_{11} + c_{1112}\varepsilon_{12} + c_{1113}\varepsilon_{13} + c_{1121}\varepsilon_{21} + c_{1122}\varepsilon_{22} + c_{1123}\varepsilon_{23} + c_{1131}\varepsilon_{31} + c_{1132}\varepsilon_{32} + c_{1133}\varepsilon_{33}$

$\sigma_1 = c_{11}\varepsilon_1 + c_{16}\varepsilon_6/2 + c_{15}\varepsilon_5/2 + c_{16}\varepsilon_6/2 + c_{12}\varepsilon_2 + c_{14}\varepsilon_4/2 + c_{15}\varepsilon_5/2 + c_{14}\varepsilon_4/2 + c_{13}\varepsilon_3$

$\quad = c_{11}\varepsilon_1 + c_{12}\varepsilon_2 + c_{13}\varepsilon_3 + c_{14}\varepsilon_4 + c_{15}\varepsilon_5 + c_{16}\varepsilon_6$

$\quad = c_{1j}\varepsilon_j$

$\sigma_{12} = c_{1211}\varepsilon_{11} + c_{1212}\varepsilon_{12} + c_{1213}\varepsilon_{13} + c_{1221}\varepsilon_{21} + c_{1222}\varepsilon_{22} + c_{1223}\varepsilon_{23} + c_{1231}\varepsilon_{31} + c_{1232}\varepsilon_{32} + c_{1233}\varepsilon_{33}$

$\sigma_6 = c_{61}\varepsilon_1 + c_{66}\varepsilon_6/2 + c_{65}\varepsilon_5/2 + c_{66}\varepsilon_6 + c_{62}\varepsilon_2/2 + c_{64}\varepsilon_4/2 + c_{65}\varepsilon_5/2 + c_{64}\varepsilon_4/2 + c_{63}\varepsilon_3$

$\quad = c_{61}\varepsilon_1 + c_{62}\varepsilon_2 + c_{63}\varepsilon_3 + c_{64}\varepsilon_4 + c_{65}\varepsilon_5 + c_{66}\varepsilon_6$

$\quad = c_{6j}\varepsilon_j$

(ii) Strain $\varepsilon_{ij} = s_{ijkl}\sigma_{kl}$

$\varepsilon_{11} = s_{1111}\sigma_{11} + s_{1112}\sigma_{12} + s_{1113}\sigma_{13} + s_{1121}\sigma_{21} + s_{1122}\sigma_{22} + s_{1123}\sigma_{23} + s_{1131}\sigma_{31} + s_{1132}\sigma_{32} + s_{1133}\sigma_{33}$

$\varepsilon_1 = s_{11}\sigma_1 + s_{16}\sigma_6/2 + s_{15}\sigma_5/2 + s_{16}\sigma_6/2 + s_{12}\sigma_2 + s_{14}\sigma_4/2 + s_{15}\sigma_5/2 + s_{14}\sigma_4/2 + s_{13}\sigma_3$

$\quad = s_{11}\sigma_1 + s_{12}\sigma_2 + s_{13}\sigma_3 + s_{14}\sigma_4 + s_{15}\sigma_5 + s_{16}\sigma_6$

$\quad = s_{1j}\sigma_j$

$\varepsilon_{12} = s_{1211}\sigma_{11} + s_{1212}\sigma_{12} + s_{1213}\sigma_{13} + s_{1221}\sigma_{21} + s_{1222}\sigma_{22} + s_{1223}\sigma_{23} + s_{1231}\sigma_{31} + s_{1232}\sigma_{32} + s_{1233}\sigma_{33}$

$\varepsilon_6/2 = s_{61}\sigma_1/2 + s_{66}\sigma_6/4 + s_{65}\sigma_5/4 + s_{66}\sigma_6/4 + s_{62}\sigma_2/2 + s_{64}\sigma_4/4 + s_{65}\sigma_5/4 + s_{64}\sigma_4/4 + s_{63}\sigma_3/2$

$\quad = s_{61}\sigma_1/2 + s_{62}\sigma_2/2 + s_{63}\sigma_3/2 + s_{64}\sigma_4/2 + s_{65}\sigma_5/2 + s_{66}\sigma_6/2$

$\quad = s_{6j}\sigma_j/2$

apply σ_1 first followed by σ_2 (case 1) and σ_2 first followed by σ_1 (case 2).

Case 1

We apply σ_1 starting from zero magnitude and gradually increase it keeping all other σs zero. Then the strain energy set up will be W_1

$$= \int_0^{\sigma_1} s_{11}\sigma_1 d\sigma_1$$

$$= s_{11}\sigma_1^2/2.$$

We now apply stress σ_2, gradually increasing it from zero upwards such that we get a further increase in strain energy of W_2

$$= \int_0^{\sigma_2} s_{22}\sigma_2 d\sigma_2 + \int_0^{\sigma_2} s_{21}\sigma_1 d\sigma_2$$

$$= s_{22}\sigma_2^2/2 + s_{21}\sigma_1\sigma_2.$$

The overall change of strain energy is then

$$W = W_1 + W_2.$$

Case 2

This time we apply σ_2 first to get a term $s_{22}\sigma_2^2/2$. Then we apply stress σ_1 and get two terms $s_{11}\sigma_1^2/2$ and $s_{12}\sigma_1\sigma_2$. Adding the three terms we again get total strain energy W.

The consequence of c_{mn} being the same as c_{nm}, and s_{mn} being the same as s_{nm}, is that there are a maximum of 21 components prior to incorporating specific crystal symmetries into the argument.

4.8 Effect of Crystal Symmetry; Equating Components by Inspection

Although we have brought the number of independent coefficients for stiffness and compliance down to 21, this is still far in excess of the number of independent components for most crystals. For cases of crystals in the hexagonal and trigonal systems it is necessary to use an analytical approach but for others we can use the method of inspection. In this method, we carry through symmetry operations associated with the crystal class and then equate the components before and after; the stiffness and compliance of the crystal must have remained unchanged as a consequence of the symmetry operation. A good example for showing the method in action is the tetragonal crystal class $\bar{4}$. This is the minimum symmetry which can be associated with the tetragonal structure. We will follow convention and take the $\bar{4}$ axis parallel to x_3. Figure 4.5 shows axes for the tetragonal crystal with the new orientations of the axes when, first, we carry out a fourfold (i.e. 90°) rotation about the x_3 axis and then carry out an inversion of the axes.

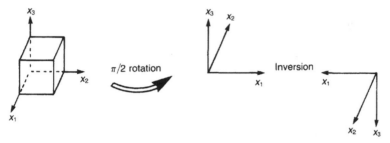

Figure 4.5 Carrying out a $\bar{4}$ operation on a tetragonal crystal.

Comparing axes before and after these operations have been achieved, we see the that the axes transform as

$$1 \rightarrow -2 \quad 2 \rightarrow 1 \quad 3 \rightarrow -3$$

where we are using 1 for axis x_1, 2 for axis x_2, 3 for axis x_3 and a negative sign to refer to an axis in the negative direction. Therefore, pairs of suffixes used to describe *tensor* components defined relative to the axes will transform as

$$11 \rightarrow 22 \quad 22 \rightarrow 11 \quad 33 \rightarrow 33$$
$$23 \rightarrow -13 \quad 31 \rightarrow 32 \quad 12 \rightarrow -21.$$

In the two-suffix description used for the *matrix* notation, we have

$$1 \rightarrow 2 \quad 2 \rightarrow 1 \quad 3 \rightarrow 3 \quad 4 \rightarrow -5 \quad 5 \rightarrow 4 \quad 6 \rightarrow -6.$$

Carrying out the transformation on the remaining 21 components we get

$$
\begin{array}{c}
\text{before} \\
\begin{pmatrix}
11 & 12 & 13 & 14 & 15 & 16 \\
   & 22 & 23 & 24 & 25 & 26 \\
   &   & 33 & 34 & 35 & 36 \\
   &   &   & 44 & 45 & 46 \\
   &   &   &   & 55 & 56 \\
   &   &   &   &   & 66
\end{pmatrix}
\end{array}
$$

$$
\begin{array}{c}
\text{after} \\
\begin{pmatrix}
22 & 21 & 23 & -25 & 24 & -26 \\
   & 11 & 13 & -15 & 14 & -16 \\
   &   & 33 & -35 & 34 & -36 \\
   &   &   & 55 & -54 & 56 \\
   &   &   &   & 44 & -46 \\
   &   &   &   &   & 66
\end{pmatrix}.
\end{array}
$$

Equating component by component we get

$$11 = 22 \quad 12 = 21 \quad 23 = 13 \quad 16 = -26 \quad 44 = 55$$

also $14 = -25$ and $25 = 14$ so $14 = 25 = 0$. Similarly, $34 = 35 = 0$ and $46 = 56 = 0$ but $16 = -26$. This leaves seven components in all and these can be represented by the following pattern :

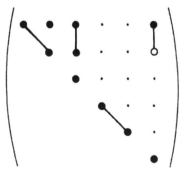

The symbols ●——○ mean components of equal magnitude but opposite sign.

As a different example consider monoclinic class 2. The twofold rotation axis (diad) is taken parallel to x_3 and there is no other symmetry operation.

Axes transform as

$$1 \to -1 \qquad 2 \to -2 \qquad 3 \to 3$$

$$11 \to 11 \qquad 22 \to 22 \qquad 33 \to 33$$

$$23 \to -23 \qquad 31 \to -31 \qquad 12 \to 12.$$

In the two-suffix notation we have

$$1 \to 1 \qquad 2 \to 2 \qquad 3 \to 3 \qquad 4 \to -4$$

$$5 \to -5 \qquad 6 \to 6.$$

Carrying out the transformation on the remaining 21 components we get

before

$$\begin{pmatrix} 11 & 12 & 13 & 14 & 15 & 16 \\ & 22 & 23 & 24 & 25 & 26 \\ & & 33 & 34 & 35 & 36 \\ & & & 44 & 45 & 46 \\ & & & & 55 & 56 \\ & & & & & 66 \end{pmatrix}$$

after

$$\begin{pmatrix} 11 & 12 & 13 & -14 & -15 & 16 \\ & 22 & 23 & -24 & -25 & 26 \\ & & 33 & -34 & -35 & 36 \\ & & & 44 & 45 & -46 \\ & & & & 55 & -56 \\ & & & & & 66 \end{pmatrix}.$$

Equating component by component we get

$$14 = -14 = 0 \qquad 15 = -15 = 0$$
$$24 = -24 = 0 \qquad 25 = -25 = 0$$
$$34 = -34 = 0 \qquad 35 = -35 = 0$$
$$46 = -46 = 0 \qquad 56 = -56 = 0.$$

We have 13 components remaining to produce the pattern

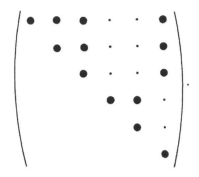

4.9 Elasticity Components in Cubic Crystals and Polycrystalline Samples

Cubic crystals are often thought to be isotropic, meaning that their properties are the same in all directions. However, this is not true for single-crystal properties that are described by higher rank tensors than second rank. The optical description 'anaxial' indicates that there is no preferred axial direction but is largely used for classifying optical samples when considering properties described by second-rank tensors. We can show that the number of independent stiffness or compliance components for a cubic single crystal is different to that for a polycrystalline sample with random orientation of the crystallites. (An amorphous sample with no long-range order of the atoms will be equivalent to a polycrystalline sample in this context.) This demonstrates that 'cubic' does not imply isotropic, except for a limited range of physical properties.

The number of independent components of compliance (or stiffness) for a cubic material can be obtained by inspection. This should be done for a particular point group within the crystal system rather than for the cubic system generally. Remember that the underlying symmetry of the cubic system is a 120° (threefold) rotation and not as is sometimes mistakenly thought a 90° rotation (see table A1 in Appendix 1). Carrying out a threefold rotation plus at least one further symmetry operation (e.g. 2 or m) will show that there are three independent components of compliance for a cubic crystal, namely s_{11}, s_{12} and s_{44} in matrix notation, and this is illustrated by the pattern

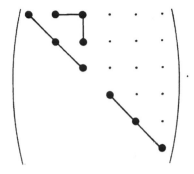

To reduce the number of components further for the isotropic (polycrystalline) case we can carry out one suitable further symmetry operation. The easiest and most obvious is to carry out a 45° rotation and equate components before and after the rotation.

Figure 4.6 shows a set of Cartesian axes x_1, x_2 and x_3 and a second broken set rotated clockwise by 45° about x_3 looking along x_3 in the positive direction. This gives direction cosines

$$l_{11} = l_{12} = l_{22}$$
$$= 1/\sqrt{2}$$

and

$$l_{21} = -1/\sqrt{2}.$$

The only components which will be involved in the transformation will be: tensor form—s_{1111}, s_{1122}, s_{2222} and all terms of type s_{1212}; matrix form—s_{11}, s_{12}, s_{22} and all terms of type s_{66}.

First we find s'_{11} in terms of components related to the old axes

$$s'_{11} = s'_{1111}$$
$$= s_{1111}/4 + s_{1122}/4 + s_{2211}/4 + s_{2222}/4 + (s_{1212}/4) \times 4$$
$$= s_{11}/4 + s_{12}/4 + s_{22}/4 + s_{21}/4 + s_{66}/4$$
$$= s_{11}/2 + s_{12}/2 + s_{66}/4$$

$$s'_{12} = s'_{1122}$$
$$= s_{1111}/4 + s_{1122}/4 + s_{2211}/4 + s_{2222}/4 - (s_{1212}/4) \times 4$$
$$= s_{11}/2 + s_{12}/2 - s_{66}/4.$$

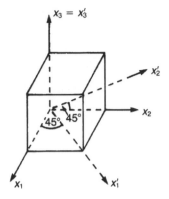

Figure 4.6 Carrying out a 45° rotation on a polycrystalline sample.

Hence

$$s'_{11} - s'_{12} = s_{66}/2$$

or

$$s_{66} = 2(s'_{11} - s'_{12}) = 2(s_{11} - s_{12}).$$

Although this equality does not reduce the filled-in spaces on the component diagram, it removes the independence of s_{66} from the other two components.

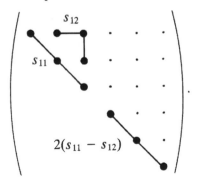

4.10 Elasticity Components in Other Crystal Systems

The method of inspection can be applied to other crystal classes. The method depends on the correspondence between orthogonal tensor components and coordinate products. Operations of the symmetry elements transform each axis either into one of the other axes or into itself in either a positive or negative direction, but overall not merely into linear combinations of themselves. Hence the feasibility of using the inspection method already described for finding the number of independent coefficients. In trigonal and hexagonal crystals there are problems. Analytical methods can be used similar to those already used in §4.9, and this involves direction cosines of 60° and 120° instead of 45°. Inspection methods can be used for hexagonal and trigonal crystals, except class 3, by the use of unusual frames of reference (Bhagavantam, 1966; Nye, reprinted version 1985).

In textbooks on elasticity there are a large number of equations relating stress and strain components for different crystal systems and for isotropic materials. Most of the relationships arise from the symmetry arguments outlined so far. However, there are further constraints. The fact that strain energy in a crystal must be positive can impose conditions on the relative magnitudes of certain of the compliance and stiffness components. These detailed relationships are beyond the scope of this book but can be pursued further by reference, for example, to Nye (1985).

4.11 Worked Examples on Stress, Strain and Elasticity

Question 4.A   Symmetric and antisymmetric tensors

Rewrite the following tensor as the sum of a symmetric and an antisymmetric tensor:

$$\begin{bmatrix} 18 & 8 & 5 \\ 6 & 2 & 4 \\ 11 & 12 & 7 \end{bmatrix}.$$

Answer

Using equation (4.4), we obtain

$$\begin{bmatrix} 18 & 8 & 5 \\ 6 & 2 & 4 \\ 11 & 12 & 7 \end{bmatrix} = \begin{bmatrix} 18 & 7 & 8 \\ 7 & 2 & 8 \\ 8 & 8 & 7 \end{bmatrix} + \begin{bmatrix} 0 & 1 & -3 \\ -1 & 0 & -4 \\ 3 & 4 & 0 \end{bmatrix}.$$

Question 4.B   Stress and strain components

A crystal sample is cut as a rectangular parallelepiped with dimensions 3 cm × 2 cm × 0.5 cm. It is compressed by a force of 150 N exerted between the two largest faces and with a force of 75 N between the 2 cm × 0.5 cm faces. As a consequence there is a reduction of thickness of 5 μm, an increase in width of 2 μm and an increase of length of 3 μm, but the sample remains a rectangular parallelepiped. Obtain the components of stress and strain for directions parallel to the sample edges.

Answer

Let x_1 be parallel to the 3 cm edge, x_2 to the 2 cm edge and x_3 to the 0.5 cm edge.

$$\sigma_1 = 150/(2 \times 0.5 \times 10^{-4}) = 1.5 \times 10^6 \text{ N m}^{-2}$$
$$\sigma_2 = 0$$
$$\sigma_3 = 75/(3 \times 2 \times 10^{-4}) = 1.25 \times 10^5 \text{ N m}^{-2}$$
$$\varepsilon_1 = 3 \times 10^{-6}/(3 \times 10^{-2}) = 10^{-4}$$
$$\varepsilon_2 = 2 \times 10^{-6}/(2 \times 10^{-2}) = 10^{-4}$$
$$\varepsilon_3 = -5 \times 10^{-6}/(5 \times 10^{-3}) = -10^{-3}$$

Problems

4.1   Rewrite the following tensor as the sum of a symmetric and an antisymmetric tensor.

$$\begin{bmatrix} 12 & 4 & 3 \\ 8 & 7 & 6 \\ 5 & 4 & 2 \end{bmatrix}$$

4.2   Show that there are nine independent components of compliance for an orthorhombic crystal of class mm2. (The crystal has two mirror planes intersecting at right angles and they intersect along a twofold rotational axis assumed to be x_3. This problem can be solved by inspection.)

4.3   Show that there are five independent components of compliance for a hexagonal crystal of class 6/m. (The crystal has a sixfold rotational axis, taken as x_3, perpendicular to a mirror plane. This problem requires an analytical approach.)

5 Crystal Optics

5.1 Introduction

Another example of a second-rank tensor is the dielectric constant for a crystalline material. The refractive index for the crystal will vary with direction. In an isotropic medium it is equal to the square root of the dielectric constant whereas in a crystalline material the principal values of the refractive index will equal the square roots of the principal components of the dielectric constant. Furthermore, care is needed as the square root relationship means that the refractive index cannot have a tensor representation.

Although we are returning to second-rank tensors in this chapter, we shall see that crystal optics involves a more complex situation than conductivity processes. The passage of light through crystals involves the propagation of transverse electromagnetic waves. Unless we are using polarised light, the wave vibrations (i.e. the electric and magnetic vectors) will be in any direction orthogonal to the direction of wave propagation. If we take the direction of wave propagation to be x, then for any one single wave, if the electric field vector is varying in the y direction, the magnetic field vector will be varying in the z direction. For all the waves passing through the crystal, the electric and magnetic vector can be in any direction in the yz plane. However, taking the electric vector as the one defining the wave vibration, the electric vectors for all the waves can be resolved into components in the orthogonal y and z directions.

If the properties of a crystal are varying markedly with orientation, then the component of the electric vector of the wave in the y direction will interact with different strength with the atoms of the crystal compared with the interaction of the z component of the electric vector with the atoms. Hence, the two components will be slowed down differently. Each component will have its own velocity in the crystal, and the crystal exhibits so-called *birefringence* or *double refraction*. To make a complete study of what will happen, we need to know not only the variation with orientation of the optical properties of the crystal, but also the direction of the light through the crystal and whether the light is unpolarised or polarised. In addition, whereas a crystal may not show birefringence in its natural state, it may become birefringent when subjected to external forces. These forces may arise from stress giving rise to the photoelastic effect, may be due to the application of an electric field giving rise to electro-optic effects, or due to a magnetic field giving rise to magneto-optic effects.

5.2 The Indicatrix

For an isotropic medium we can represent the dielectric properties at optical frequencies by the equation

$$D = \varepsilon_0 K E = \varepsilon E$$

where E is electric field, D is electric displacement (electric flux density) and ε_0 is the permittivity of free space. K is the dielectric constant and ε the permittivity of the medium.

For an optical medium which is not isotropic we can write

$$D_i = \varepsilon_0 K_{ij} E_j = \varepsilon_{ij} E_j. \tag{5.1}$$

Both K_{ij} and ε_{ij} are symmetrical second-rank tensors and can be represented in similar ways to the thermal conductivity tensor k_{ij} and the thermal resistivity tensor r_{ij} by representation quadrics. We will use the ε_{ij} form of equation (5.1). Representation by K_{ij} would be entirely similar but with ε_0 acting as a scaling factor. (Using K_{ij} would have the advantage of keeping the numerical values close to unity.)

Equation (5.1) involving ε_{ij} is analogous to the equation for heat flow and the permittivity tensor is analogous to the conductivity tensor. Just as we also used a resistivity tensor, so we can also use a reciprocal dielectric tensor $(K^{-1})_{ij}$ which we will represent by the symbol η_{ij}. This tensor is sometimes called the optical impermeability. Figure 5.1 makes the comparison between the ellipsoids for conductivity and permittivity and the ellipsoids for resistivity and the reciprocal dielectric tensor. For the latter, the ellipsoid is given the special name of optical *indicatrix*. We can write

$$n_1 = \sqrt{K_1} = 1/\sqrt{\eta_1}$$
$$n_2 = \sqrt{K_2} = 1/\sqrt{\eta_2} \qquad (5.2)$$
$$n_3 = \sqrt{K_3} = 1/\sqrt{\eta_3}$$

where n_1, n_2 and n_3 are the principal values of the refractive index. If x_1, x_2 and x_3 correspond to the principal axes of the representational ellipsoid then we can write

$$x_1^2/n_1^2 + x_2^2/n_2^2 + x_3^2/n_3^2 = 1. \qquad (5.3)$$

This is defining the indicatrix in terms of the principal values of the refractive index. For a cubic crystal, $n_1 = n_2 = n_3$ and the indicatrix is a sphere. There will be no double refraction in a crystal having cubic symmetry.

For tetragonal, hexagonal and trigonal crystals the indicatrix is an ellipsoid of revolution about the principal symmetry axis. This principal axis is called the *optic axis* and the crystals are said to be *uniaxial*. The refractive index in the direction of this axis is $n_3 = n_E$. We use the notation n_E to represent the so-called *extraordinary* refractive index. The values of the refractive index for the other axes of the ellipsoid are equal: $n_1 = n_2 = n_O$ where n_O represents the so-called *ordinary* refractive index.

For orthorhombic, monoclinic and triclinic crystal systems, the indicatrix is a triaxial ellipsoid. The reader may wish to refer back to table 2.1 where details of the ellipsoids for the different crystal systems are given. There are *two* circular cross sections and hence two privileged wavevector directions perpendicular to these circular cross sections for which there is no double

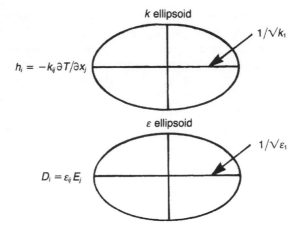

$$h_i = -k_{ij}\partial T/\partial x_j$$

$$D_i = \varepsilon_{ij}E_j$$

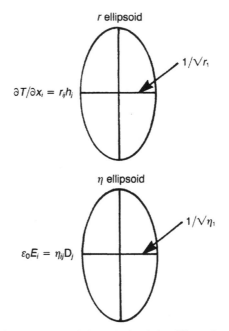

$$\partial T/\partial x_i = r_{ij}h_j$$

$$\varepsilon_0 E_i = \eta_{ij}D_j$$

Figure 5.1 Comparison of the conductivity (k) and permittivity (ε) ellipsoids and the resistivity (r) ellipsoid and the optical indicatrix (η ellipsoid).

refraction (figure 5.2). These two directions perpendicular to the circular cross sections are called (primary) optic axes and because there are these two directions in which no birefringence is observed, crystals of this type are called *biaxial*.

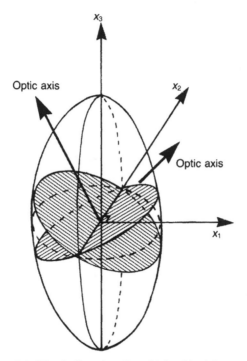

Figure 5.2 The indicatrix ellipsoid for biaxial crystals.

If the reader does not find it obvious that there will be two circular cross sections to the ellipsoid in figure 5.2, consider the cross section in the x_2x_3 plane with the x_3 axis longer than the x_2 axis. It is conventional to define the ellipsoid with the relative lengths of the axes as shown corresponding to $n_3 > n_2 > n_1$. Now rotate this cross section towards x_1. What starts off as an elongated ellipse must become a flattened ellipse by the time it is in the x_1x_2 plane. Somewhere between the two extremes x_1x_3 and x_1x_2 there must be an orientation in which the cross section is no longer elongated in either direction and is circular. The second circular cross section can be found by rotating towards the x_1x_2 plane but this time in the opposite sense towards $-x_1$.

5.3 The Wave Surface

For optical propagation in crystals, it is common to show the wave surface diagrammatically. The surface is different to both the indicatrix and (as we shall see) the wavevector surface, although they are all directly related. Consider a uniaxial crystal. Suppose a point source of light is situated within the crystal. The wavefront emitted at any instant of time forms a continuously expanding surface. If we fix this surface for some subsequent instant in time, the surface forms the so-called wave surface. The situation is somewhat similar to considering thermal flow from a point source of heat. The wave surface is analogous to an isothermal surface. However, there is an important difference. In general, two rays propagate from the light source with different velocities in any direction. (Remember that the light can have an electric vector which can lie in any direction in the plane orthogonal to the wavevector or wave direction.) Resolved components can correspond to refractive indices n_O and n_E, thus giving rise to two velocities. The waves will move outwards with two different velocities. These two velocities give rise to a double wavefront surface and we usually say that the surface has two 'sheets' to it (see figure 5.3). In the direction of the optic axis the sheets will touch. Any electric vector for a wave moving along the optic axis will be vibrating in the plane of circular cross section within the indicatrix; all the waves will proceed with the same velocity. The ordinary sheet of the wave surface will be a sphere whose radius is proportional to $1/n_O$, whereas it can be shown by geometry that the extraordinary sheet of the wave surface is an ellipsoid of revolution about the optic axis with the lengths of the semiaxes proportional to $1/n_E$, $1/n_O$ and $1/n_O$.

It is a confusing feature of crystal optics that in addition to the wave surface, it is also common to show the wavevector surface. We have seen that wave surfaces measure the distances reached by light from a point source for different polarisations and hence become large for large light velocities in the crystal. On the other hand, wavevector surfaces measure the variations of k value with orientation for different light polarisations. As $k = 2\pi/\lambda = \omega n/c$, where λ = wavelength, ω = angular frequency and c = velocity of light in free space, wavevector surfaces become smaller as the velocity of light in the crystal increases. (n is smaller for larger light velocity in the crystal.) Because of the direct proportionality between k and n, the wavevector surfaces are also called n-surfaces but they *should not be confused* with the indicatrix ellipsoid. Figure 5.4 shows the sheets both of the wave surface and of the wavevector surface to emphasise the inverse relationship. The figure also shows the forms of the waves (the variation of the electric components) as they pass through the crystal with the k vector orthogonal to the optic axis. The two components of the electric vector correspond to refractive indices n_O and n_E and hence we have wavelengths $2\pi c/n_O\omega$ and $2\pi c/n_E\omega$ for the light within the crystal.

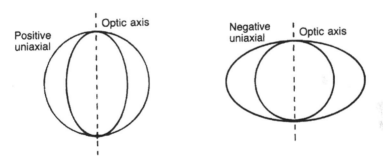

Figure 5.3 The wave surfaces for uniaxial crystals.

Another term which is sometimes used is that of wave normal. It refers to the direction which is normal to the wavefront and so is normal to the tangential plane at a point on the wave surface. It will not be in the same direction as the wave direction unless the wave is moving along a principal axis (figure 5.5).

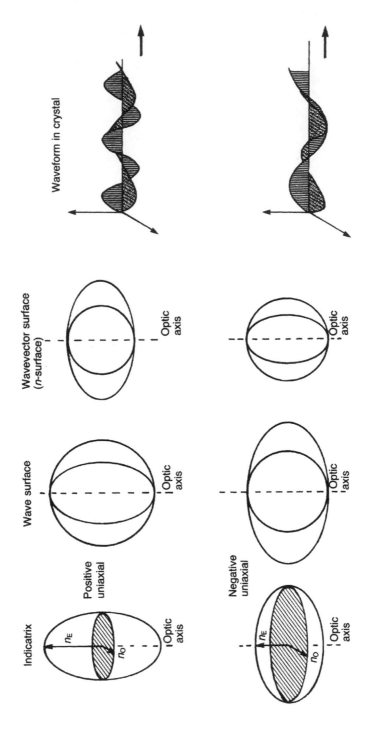

Figure 5.4 The relationship between the indicatrix, the wave surface, the wavevector surface and the waveform of the emerging light waves in positive and negative uniaxial crystals.

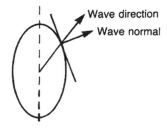

Figure 5.5 Wave direction and wave normal.

5.4 Biaxial Crystals

For biaxial crystals the situation is less simple! Consider a wave moving out along any one of the three principal directions x_1, x_2 or x_3. The waves will have wavevectors k_1, k_2 and k_3. The electric vectors for these waves lie within the elliptical cross sections of the indicatrix and so there will be two possible principal refractive indices, and hence two possible velocities for the light. For instance in the x_1 direction we will have velocities c/n_2 and c/n_3. These will give values of k_1 of

$$k_1 = n_2\omega/c \quad \text{and} \quad k_1 = n_3\omega/c.$$

If we refer back to the particular indicatrix illustrated in figure 5.2, the value of n_3 is larger than n_2 and gives the larger value of k_1. As mentioned, it is conventional to take $n_1 < n_2 < n_3$.

Similarly we will have values for k_2 of

$$k_2 = n_1\omega/c \quad \text{and} \quad k_2 = n_3\omega/c.$$

We can now plot cross sections for the two sheets of the wavevector surface for the k_1k_2 principal plane and these are shown in figure 5.6. The sheet involving n_3 is circular in the k_1k_2 plane. (In whichever direction k lies within the x_1x_2 plane, there always will be a possible electric vector component in the x_3 direction.) The cross section of the other sheet is elliptical as a consequence of taking into account the different values of the wavevector for the two principal axes and all the intermediate directions, and hence intermediate k values.

Similarly we can plot cross sections for the wavevector sheets for the k_2k_3 principal plane and the k_1k_3 principal plane. The

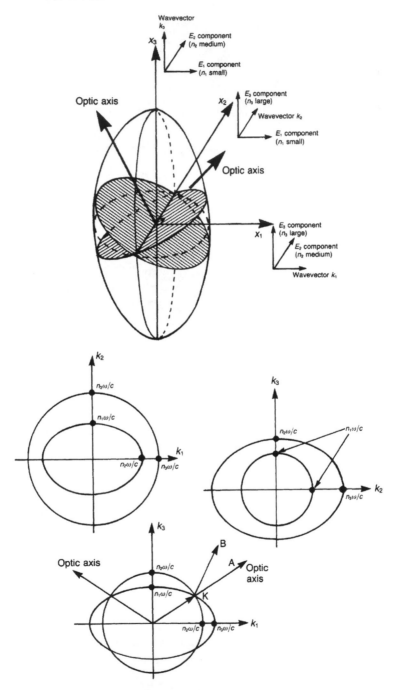

wavevector sheets within the $k_1 k_3$ plane cross with a common value for k. This occurs in the directions of the optic axes where the k vector is perpendicular to the circular cross sections of the indicatrix. It is worth noting that at the point of intersection, although the k values are the same, the wavefronts are moving out in different directions perpendicular to the respective k surfaces. Referring to the $k_1 k_3$ cross section illustrated in figure 5.6, KA represents the direction of movement of the wavefront for an x_2-polarised wave and KB repesents the direction of movement of the wavefront for a wave polarised in the $x_1 x_3$ plane.

To describe the wavevector surface completely we need to show the overall surface in three dimensions and this surface will consist of two complete sheets. The cross sections shown so far are the principal cross sections for these sheets, and by thinking of these two sheets as continuous surfaces one can derive the three-dimensional configuration for the wavevector surface as shown in figure 5.7.

5.5 Double Refraction (Birefringence) at a Boundary

So far we have considered only what happens to the rays of light as they move through the crystal itself. If we consider a plane wave incident on the surface of a crystal at an angle of incidence other than zero (whereas we usually consider a plane wave incident normal to the crystal surface) then the wave will be refracted. If the wave is incident onto a *uniaxial* crystal, the refracted wave will be in general a mixture of the ordinary and the extraordinary wave. The wavevectors must lie in the plane of incidence and the tangential components along the air–crystal boundary must be in the same plane. The wave can be broken down into two components each meeting the condition of conservation of the wavevector component parallel to the interface. Thus

$$k_a \sin \theta_a = k_O \sin \theta_O = k_E \sin \theta_E \qquad (5.4)$$

Figure 5.6 Cross sections across the principal planes of the wavevector sheets for a biaxial crystal whose indicatrix is as shown in figure 5.2.

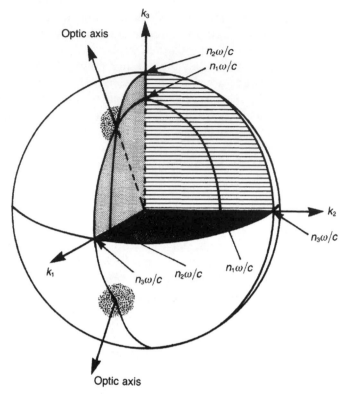

Figure 5.7 The wavevector surface in three dimensions for a biaxial crystal.

where k_a is the wavevector for the incident wave in air and k_O and k_E are the wavevectors for the ordinary and extraordinary wave components, respectively, in the crystal. θ_a is the angle of incidence of the incoming wave in air and θ_O and θ_E are the angles of refraction for the refracted waves having wavevectors k_O and k_E respectively (see figure 5.8(*a*)). The ordinary sheet of the wavevector surface for a uniaxial crystal is a sphere so k_O is fixed in magnitude and direction by Snell's law: i.e.

$$n_a \sin \theta_a = n_O \sin \theta_O. \tag{5.5}$$

Thus θ_O is fixed. The extraordinary sheet is an ellipsoid of revolution (but it has a circular cross section perpendicular to the optic axis) so k_E varies with direction. However, there will

be only one magnitude and direction of k_E which will satisfy the
requirements of both the wavevector surface and of Snell's law.

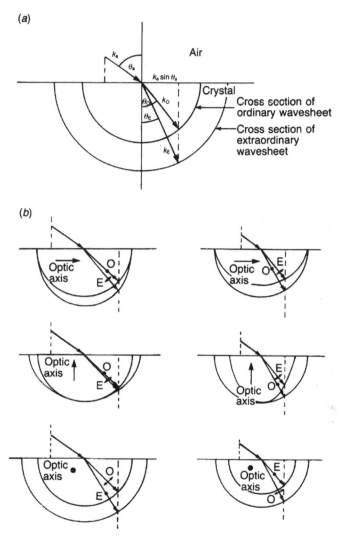

Figure 5.8 (*a*) Double refraction at a boundary. (*b*) Examples in positive and negative uniaxial crystals.

Algebraically, the values of k_E and θ_E are solved from a quartic
equation. A graphical solution is easier as shown by figure

5.8(a) where all wavevectors lie in the plane of the paper. This figure corresponds to taking a positive uniaxial crystal with the optic axis parallel to the boundary and perpendicular to the plane of incidence. Hence the plane of the paper lies within the circular cross section of the extraordinary sheet and the cross section of the ordinary sheet must be circular. Having fixed k_O (magnitude and direction), we can also fix k_E in magnitude and direction. From equation (5.4), k_O and k_E have the same resolved component $k_a \sin \theta$ parallel to the interface and this is illustrated geometrically. Figure 5.8(b) shows further examples of double refraction which are possible in positive uniaxial and negative uniaxial crystals. Arrows and dots on the k vectors show whether the ordinary and extraordinary waves have electric vectors in the plane or perpendicular to the plane of incidence respectively. The direction of the optic axis in each case is shown by an arrow or alternatively by a dot to indicate that it has a direction perpendicular to the plane of the paper. Comparison of the illustrations in figure 5.8(b) with those for the *wavevector surfaces* in figure 5.4 can be made by lining up the optic axes correctly.

5.6 Worked Examples on Polarisation and Birefringence

Question 5.A Birefringence and quarter-wave plate
A linearly polarised light beam with $\lambda = 589.3$ nm is incident normally onto the yz face of a quartz plate. (Such a plate is said to be x-cut and the z axis corresponds to the hexagonal c axis.) The wave is incident to the face at $x = 0$ and passes through the crystal along the x axis. It is initially polarised such that it has components of equal amplitude in the y and z directions. What is the state of polarisation at position x where $(k_z - k_y)x = \pi/2$ and how does the electric vector vary with time? (A plate of such a thickness x is called a quarter-wave plate.)

Determine the value of x assuming $n_O = 1.544$ and $n_E = 1.553$ for quartz.

Answer
The light is circularly polarised with the electric vector rotating with time.

$$(2\pi/\lambda_z - 2\pi/\lambda_y)x = \pi/2$$

$$\lambda_z = \frac{589.3}{1.553}$$

$$\lambda_y = \frac{589.3}{1.544}$$

$$\frac{2\pi}{589.3}(1.553 - 1.544)x = \pi/2$$

$$x = 16.4 \; \mu\text{m}$$

Question 5.B   Elliptically polarised light and birefringence

Describe the state of polarisation of a wave represented by

$$E_y = A \cos(\omega t - kx) \qquad\qquad (5.6a)$$

$$E_z = 2A \cos(\omega t - kx + \pi/4) \qquad\qquad (5.6b)$$

where A is a constant.

The wave passes normally through a quarter-wave plate (see Question 5.A). What is the state of polarisation of the emergent wave, assuming the primary refractive indices lie in the y and z directions and that the larger value lies in (i) the z direction, (ii) the y direction?

Answer

The z component leads the y component by $\pi/4$ and has twice the amplitude of the y component. This gives a polarisation ellipse as shown in figure 5.9(a). The ellipse can be obtained graphically by taking ωt to vary between 0 and 2π and plotting z against y, or it can be obtained algebraically by eliminating $(\omega t - kx)$ from equations (5.6a) and (5.6b).

(i) When the light passes through the quarter-wave plate with the larger refractive index in the z direction, $(k_z - k_y)x = \pi/2$, assuming we take the origin of x at the initial face of the plate. When the wave emerges at the opposite face

$$E_y = A \cos(\omega t - k_y x)$$

$$E_z = 2A \cos(\omega t - k_y x - \pi/2 + \pi/4)$$

$$= 2A \cos(\omega t - k_y x - \pi/4).$$

The light remains elliptically polarised and similarly oriented to that entering the crystal but the direction of rotation is reversed (figure 5.9(b)).

(ii) If the smaller refractive index lies in the z direction $(k_y - k_z)x = \pi/2$

$$E_y = A\cos(\omega t - k_y x)$$
$$E_z = 2A\cos(\omega t - k_y x + 3\pi/4).$$

The polarisation ellipsoid has rotated by $\pi/2$ as shown in figure 5.9(c).

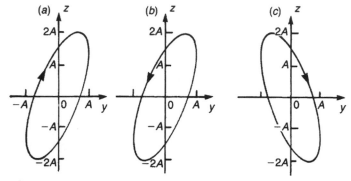

Figure 5.9 Polarisation states for light wave in Question 5.B. (a) Light on entering $\lambda/4$-wave plate. (b) Light emerging from $\lambda/4$-wave plate when $n_z > n_y$. (c) Light emerging from $\lambda/4$-wave plate when $n_z < n_y$.

Problems

5.1 Mica cleaves naturally on the (001) plane such that it is birefringent. Cleaved to an appropriate thickness it can act as a quarter-wave plate. If the principal refractive indices are $n_1 = 1.552$, $n_2 = 1.582$ and $n_3 = 1.588$, calculate the required thickness of mica when light from a He–Ne laser of wavelength in air of 633 nm is incident perpendicularly onto the surface.

5.2 A half-wave plate is cut from calcite, which has rhombohedral crystal structure but for which a hexagonal cell can be used. Calculate the wave-plate thickness for a plate cut such that (i) the hexagonal c axis lies in the plane of incidence, (ii) the c axis is at $45°$ to the plane of incidence. Take $n_O = 1.6584$, $n_E = 1.4864$ and wavelength 633 nm. In practice one would use configuration (i).

5.3 (*a*) Waves entering a quarter-wave plate are represented in three different cases by

(i) $E_y = A \cos(\omega t - kx)$ $E_z = A \sin(\omega t - kx)$
(ii) $E_y = \sqrt{3}A \cos(\omega t - kx)$ $E_z = A \cos(\omega t - kx)$
(iii) $E_y = 3A \cos(\omega t - kx)$ $E_z = A \cos(\omega t - kx - 3\pi/4)$.

Describe the polarisation state for the wave in each case.

(*b*) Give the polarisation states for the waves when they emerge from the wave plate assuming that the two principal refractive indices lie in the y and z directions. Where appropriate, differentiate between the two cases of the y components of the waves being either faster or slower than the z components.

5.4 Antimony thioiodide (SbSI) crystals are biaxial. The principal refractive indices exhibit large variations: $n_1 = 2.7$, $n_2 = 3.2$ and $n_3 = 3.8$. Plot the cross section within the $k_1 k_3$ principal plane of the wavevector sheets. Hence deduce the direction of the optic axis and check the value by calculation.

6 Axial Tensors

6.1 Definition of an Axial Tensor

On p.26 we defined a second-rank tensor representing a physical property as a quantity which transforms according to

$$T'_{ij} = l_{ip}l_{jq}T_{pq}. \tag{6.1}$$

However, there are other properties such as optical activity (rotation of the plane of polarisation of light) which transform according to

$$T'_{ij} = \pm l_{ip}l_{jq}T_{pq}. \tag{6.2}$$

The plus and minus in optical activity are required to indicate the direction of rotation. We shall see that there are a number of physical properties which transform according to equation (6.2) and these are referred to as pseudo-tensors or axial tensors. Axial vectors transform similarly by

$$r'_i = \pm l_{ij}r_j \tag{6.3}$$

and we can go on to define higher-rank axial tensors also. If it is necessary to make explicit distinction between the pseudo-tensors (i.e. axial tensors) and the real tensors, then the real tensors are called *polar tensors*.

6.2 Transformation of Axial Vectors

Because an axial vector involves an additional piece of information, i.e. the sense of the rotation, the overall transformation does not depend only on the transformation of the axes. We need to look at the relationships between polar and axial vectors for different symmetry axes, different symmetry planes and inversion through a centre of symmetry. These cases are illustrated in figure 6.1.

Figure 6.1(a) illustrates the case of a twofold rotational axis of symmetry defined as x_3. For a polar vector, carrying through the 180° rotation results in x_1, x_2 and x_3 becoming $-x_1$, $-x_2$ and x_3. Carrying out a similar operation on an axial vector, we have a rotation ABC (clockwise rotation looking along OP) becoming a rotation A'B'C' (clockwise direction looking along OP') and the axial transformation is also equivalent to

$$x_1, x_2, x_3 \rightarrow -x_1, -x_2, x_3.$$

Hence a diad axis acts on an axial vector in the same way as it does on a polar vector. A similar study of the effects of three-, four- and sixfold axes shows that they all behave in a similar way on polar and axial vectors.

Figure 6.1(b) shows the effect of a mirror plane x_1x_2. Polar vector OP in the x_1, x_2, x_3 direction becomes a polar vector OP' in the x_1, x_2 and $-x_3$ direction. Doing the same for an axial vector OP, it first becomes OP' but clockwise rotation ABC as we look along OP becomes anticlockwise rotation A'B'C' as we look along OP'. We can reverse the rotation by looking in the opposite direction, i.e. by looking along P'O. This is equivalent to our axial tensor extending from the origin O in the direction OP". Overall, the axial vector OP represented by x_1, x_2, x_3 becomes $-x_1$, $-x_2$, x_3. Hence the axial transformation is the negative of the polar transformation.

Figure 6.1(c) illustrates a centre of symmetry. For polar vector OP we transform to OP' and x_1, x_2 and x_3 transform to $-x_1$, $-x_2$ and $-x_3$. In the case of the axial vector OP, clockwise rotation ABC becomes anticlockwise rotation A'B'C' when we transform to OP'. So looking in the reverse direction, we return our vector to OP and it becomes unchanged (x_1, x_2, x_3 unchanged). Although we have no change of sign for the axes, the transformation involves a *reversal of sign compared with the situation for a polar vector*. Summarising, we have for axial vectors

$$r_i' = l_{ij}r_j \qquad \text{for axes of rotation}$$

$$r_i' = -l_{ij}r_j \qquad \text{for planes and centres of symmetry.}$$

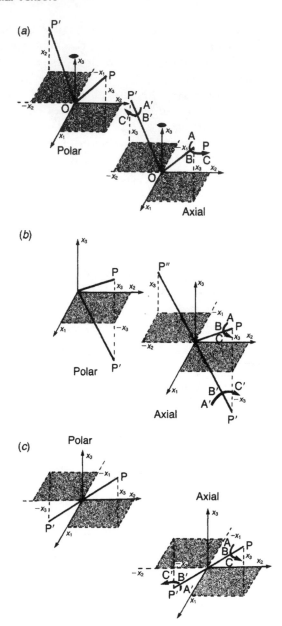

Figure 6.1 Transformation of polar and axial vectors for symmetry operations. (*a*) Twofold rotational axis, (*b*) mirror plane, (*c*) centre of symmetry.

6.3 Transformation of Axial Tensors

To transform a second-rank axial tensor T_{ij} we use the results that we have just summarised in §6.2 above. We consider the transformation of axes for the symmetry operations applicable to a particular crystal. For instance, consider a mirror plane in which the mirror is perpendicular to x_2. Axes transform from 1 to 1, 2 to -2 and 3 to 3. The direction of rotation is changed giving an additional negative in each case, and we take this into account after considering the transformation of the axes. Therefore transforming by inspection we have

$$T_{11} \to -T_{11} \qquad T_{12} \to T_{12} \qquad T_{31} \to -T_{31}$$
$$T_{22} \to -T_{22} \qquad T_{23} \to T_{23} \qquad T_{33} \to -T_{33}.$$

For a centre of symmetry, 1 goes to -1, 2 to -2 and 3 to -3 and in addition there is change of direction of rotation. So

$$T'_{ij} = -T_{ij}$$

(two negative signs for the changes of axes and an additional one for the rotational change, all multiplying to give negative overall). An example involving a second-rank axial tensor is shown on p.98 where the gyration tensor for optical activity is discussed.

Similar arguments can be extended to higher-order axial tensors.

6.4 Optical Activity

In Chapter 5 we discussed basic ideas of the optical properties of anaxial, uniaxial and biaxial crystals. The coverage considered so far does not include certain other optical properties. In particular, if a linearly polarised light beam is passed through certain optical media, it is found that the plane of polarisation is rotated, the amount of rotation being proportional to the path length of the light through the medium. The phenomenon can be found in certain anaxial and biaxial crystals even when the light is passing in the direction of the optical axis and ordinary double refraction is absent. The classic example of a crystal showing this effect, which is called *optical rotatory power* or optical activity, is α-quartz.

The sense of rotation is fixed relative to the wavevector of the light. If the light is made to traverse a crystal once forwards and then reflected backwards through the same length of crystal, the net rotation is zero. A medium is said to be laevorotatory or left handed if it rotates the plane of polarisation anticlockwise as one looks along the crystal in the same direction as the wavevector of the light, and it is said to be dextrorotatory or right handed if it rotates the light in a clockwise direction.

To explain what is happening we shall need to consider circularly polarised light. But first consider a plane-polarised beam of light whose displacement vector amplitude we can represent by

$$y = A \cos[\omega(t - xn/c)] \qquad (6.4)$$

where A is the amplitude of the displacement vector, ω is the angular frequency of the light, t is time, n is the refractive index of the medium, c is the velocity of light in free space (i.e. c/n is the velocity of the light within the crystal) and the wavevector is in direction x. This will be equivalent to

$$
\begin{aligned}
y_1 &= \tfrac{1}{2}A \cos[\omega(t - xn/c)] \\
z_1 &= \tfrac{1}{2}A \sin[\omega(t - xn/c)] \\
y_2 &= \tfrac{1}{2}A \cos[\omega(t - xn/c)] \\
z_2 &= -\tfrac{1}{2}A \sin[\omega(t - xn/c)]
\end{aligned}
\qquad (6.5)
$$

where the two contributions in the z direction (orthogonal to y) cancel and the two y contributions add to give us our earlier contribution for y in equation (6.4). Equations (6.5) represent two circularly polarised light waves of amplitude $\tfrac{1}{2}A$ rotating in opposite directions as we can see by combining as follows

$$(y_1^2 + z_1^2) = (\tfrac{1}{2}A)^2$$

and

$$(y_2^2 + z_2^2) = (\tfrac{1}{2}A)^2.$$

In an optically active crystal we associate two separate refractive indices n_l and n_r with the left-handed and right-handed circularly polarised waves. Our equations now become

$$y_1 = \tfrac{1}{2}A \cos[\omega(t - xn_1/c)]$$
$$z_1 = \tfrac{1}{2}A \sin[\omega(t - xn_1/c)]$$
$$y_2 = \tfrac{1}{2}A \cos[\omega(t - xn_r/c)]$$
$$z_2 = -\tfrac{1}{2}A \sin[\omega(t - xn_r/c)]$$

$$(6.6a)$$

or alternatively as only relative phases are important:

$$y_1 = \tfrac{1}{2}A \cos \omega t$$
$$y_2 = \tfrac{1}{2}A \cos(\omega t - \delta)$$
$$z_1 = \tfrac{1}{2}A \sin \omega t$$
$$z_2 = -\tfrac{1}{2}A \sin(\omega t - \delta)$$

$$(6.6b)$$

where

$$\delta = \omega x(n_1 - n_r)/c.$$

We use the standard trigonometric relationships

$$\cos A + \cos B = 2\cos[(A + B)/2]\cos[(A - B)/2]$$
$$\sin A - \sin B = 2\cos[(A + B)/2]\sin[(A - B)/2].$$

Combining the two y contributions and combining the two z contributions to obtain the overall amplitude in the y direction and the overall amplitude in the z direction we have

$$Y = y_1 + y_2$$
$$= A \cos(\omega t - \delta/2) \cos \delta/2$$
$$Z = z_1 + z_2$$
$$= A \cos(\omega t - \delta/2) \sin \delta/2.$$

$$(6.7)$$

The overall intensity resulting from the amplitudes in the y and z directions is given by

$$Y^2 + Z^2 = A^2 \cos^2(\omega t - \delta/2). \qquad (6.8)$$

This expression for intensity at any instant of time t is equivalent to the intensity for a linearly polarised beam rotated clockwise by an angle $\phi = \delta/2 = \omega x(n_r - n_1)/2c$ with respect to the original direction of the linear polarisation (see figure 6.2(a)).

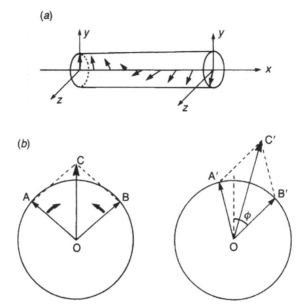

Figure 6.2 (*a*) Rotation of a linearly polarised light wave. (*b*) Combining two circularly polarised light waves before and after passing through a crystal.

Figure 6.2(*b*) shows how this operates in a crystal. Consider vectors OA and OB representing the electric vectors for circularly polarised light waves which are polarised right handed and left handed respectively. They combine to give a plane-polarised resultant OC. When the waves have passed through the crystal and reach the opposite face of the crystal, the relative angular position of OA (now OA′) and OB (now OB′) will depend on the difference between n_r and n_l. If they are as shown in the figure then they produce resultant OC′ which has rotated by an angle ϕ relative to OC. Thus, with $n_r > n_l$ the rotation shown is clockwise; the plane of polarisation turns in the same sense as the circularly polarised wave which is travelling with the greater velocity.

The rotation per unit length in the crystal is called the *specific rotatory power* of the crystal and is given conventionally by

$$\rho = \phi/d = \pi(n_l - n_r)/\lambda_0. \tag{6.9}$$

For α-quartz, $n_l - n_r = 6.6 \times 10^{-5}$ at 633 nm. Although this difference in magnitude of the two refractive indices is very small, the difference gives rise to a specific rotatory power of $18.8° \, \text{mm}^{-1}$. This can be measured accurately.

6.5 Optical Activity in the Presence of Birefringence

The arguments regarding optical activity refer to the rotation which occurs for left-handed and right-handed components of circularly polarised light. The optical activity can occur in isotropic materials and cubic crystals, and is applicable to light passing along the optic axes in less symmetrical optical crystals. Why such an effect should arise from the production of induced dipole moments on the molecules in the presence of the electric and magnetic fields associated with the light waves is beyond the scope of this text. (For further discussion see, for example, Yariv and Yeh (1984, p.96).) What is important is that this effect is additional to the birefringence effects in uniaxial and biaxial crystals. The optical activity may be considered to be a small perturbation superimposed on the ordinary birefringence.

We can draw the wave surface for α-quartz showing the two sheets, the sphere and the ellipsoid. These surfaces will represent the outwards propagation of two orthogonal plane-polarised waves. If optical activity is present these plane-polarised waves can themselves be broken down into circular components whose refractive indices are modified from the mean values giving rise to the wavevector sheets shown as broken curves in figure 6.3. Along the optic axes, whereas previously we had the two sheets touching with the presence of optical activity, we now have one circular component which is slightly speeded up and one circular component which is slightly slowed down. Similarly, we can see a corresponding speeding up and slowing down along the principal axes orthogonal to the optic axis. At intermediate orientations, the radial distortion of the wave surface is as shown (but very considerably exaggerated in the figure). The points P_1, P_2 and Q_1, Q_2 and symmetrically related points show where the undistorted and distorted sheets touch. A detailed analysis is lengthy and will not be included here.

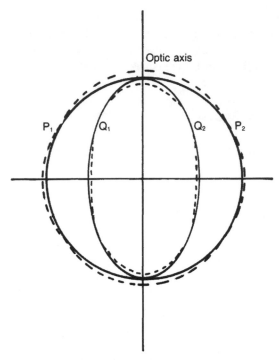

Figure 6.3 The wave surface for α-quartz showing the presence of both birefringence and optical activity.

6.6 The Gyration Tensor; Second-rank Axial Tensor

The tensor which is used to represent optical activity is called the gyration tensor and is denoted g_{ij}. It is a second-rank axial tensor and is symmetric. Thus it will transform as discussed in §6.3. In addition, we can use the applicable symmetry arguments for different crystal classes, just as we did for polar tensors. What is not immediately obvious is the relationship between the gyration tensor g_{ij} and the specific rotatory power ρ.

There is an optical activity constant G in the direction of propagation defined by

$$G = g_{ij}l_il_j \tag{6.10}$$

where subscripts $i, j = 1, 2, 3$, and the use of subscripts implies summation. It is conventional here to write the direction cosines, the ls, after the gyration tensor g_{ij}. The direction cosines l_1, l_2 and l_3 define the cosines of the angles between the direction of the wave normal, and the chosen set of orthogonal axes. Thus

$$G = g_{11}l_1^2 + g_{22}l_2^2 + g_{33}l_3^2 + 2g_{12}l_1l_2 + 2g_{23}l_2l_3 + 2g_{13}l_1l_3$$

$$(6.11)$$

as $g_{ij} = g_{ji}$. (It cannot matter as to the order in which l_i and l_j are taken and thus there is no physical difference between g_{ij} and g_{ji}.)

Also,

$$G^2 = (n^2 - n'^2)(n^2 - n''^2) \qquad (6.12)$$

where n' and n'' are the two refractive indices which the crystal would have for the particular direction of propagation in the absence of optical activity. n'^2 and n''^2 arise as roots of the so-called Fresnel equation (for further details see, for example, Yariv and Yeh (1984)). n is the refractive index in the presence of optical activity. Note that n' and n'' are not principal refractive indices and hence the use of the new (primed) notation.

It is usual to consider as a special and more simple case that of wave propagation along the optic axis when $n' = n'' = n_O$ and equation (6.12) simplifies to

$$n^2 = n_O^2 \pm G.$$

G is always small so that we can write by expansion

$$n = n_O \pm G/2n_O \qquad (6.13)$$

and this corresponds to two circularly polarised waves with rotatory power

$$\rho = \pi G/\lambda n_O \qquad (6.14)$$

by using equation (6.9) and $n_l - n_r = G/n_O$.

Hence the relationship between G and ρ becomes apparent although the origin of G has not been obtained rigorously, and we have not shown the relationship for a general direction.

Let us use our earlier example of a $\bar{4}$ operation on a tetragonal crystal and obtain the number of independent gyration tensor components. Figure 6.4 shows the transformation. The figure is identical to figure 4.5 except that additionally we now have rotational directions inserted on the axes. As before

$$1 \to -2 \quad 2 \to 1 \quad 3 \to -3$$

but the direction of rotation is reversed. Therefore we must change the sign of g_{ij} and components of g_{ij} transform as follows

$$g_{11} \to -g_{22} \quad g_{22} \to -g_{11} \quad g_{33} \to -g_{33} = 0$$

$$g_{23} \to g_{13} \to -g_{23} = 0 \quad g_{12} \to g_{21}$$

giving

$$\begin{bmatrix} g_{11} & g_{12} & 0 \\ & -g_{11} & 0 \\ & & 0 \end{bmatrix}.$$

The representational surface for optical activity for a $\bar{4}$ tetragonal crystal lies in the $x_1 x_2$ (i.e. 'aa') plane. We cannot represent the surface by a real ellipse as one of the two principal components is negative.

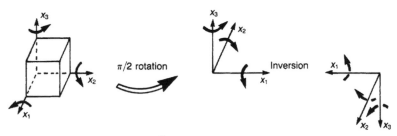

Figure 6.4 Carrying out a $\bar{4}$ operation on a tetragonal crystal with rotation included (the axial case).

Returning to crystal classes in general, carrying out a transformation through a centre of symmetry reverses the sign of g_{ij} because of reversal of the rotation. However, we have stated that $g_{ij} = g_{ji}$. As a consequence, $g_{ij} = 0$ for crystal classes possessing a centre of symmetry. This means that the following classes, which are centrosymmetric, do not show optical activity:

$\bar{1}$ 2/m   mm2   4/mmm $\bar{3}$m   6/m   6/mmm

m3   m3m.

In addition, analyses of the type shown above give all components of g_{ij} to be zero for the following crystal classes:

3m $\bar{6}$ 6mm   4mm $\bar{6}$m2 $\bar{4}$3m.

Hence these also possess no optical activity.

6.7 The Hall Effect; Third-rank Axial Tensor

In §3.4 we considered the passage of an electrical current along a long narrow sample when an electric potential is applied between the ends. To represent this situation, we used the second-rank resistivity tensor ρ_{ij}. In the case of the Hall effect, a field of magnetic induction B is applied across the sample in addition to the electric current flowing along the sample. The usual configuration is to use a sample which is a rectangular parallelepiped. The magnetic field is applied perpendicular to the larger faces (figure 6.5).

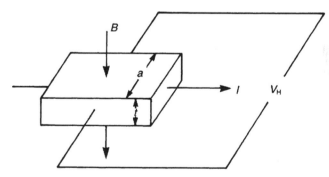

Figure 6.5 The Hall effect configuration.

The presence of the magnetic field produces a force on the moving charges within the conductor and initially there is a build-up of charge on the side faces of the conductor. This establishes a voltage V_H between the side faces, and hence an electric field component across the sample orthogonal to both the current I and the magnetic induction B. The electric field established when equilibrium is achieved is of magnitude such that the force on the charges due to the electric field balances the magnetic force on the charges. The charges pass undeflected through the conductor. However, the charges will possess a range of velocities so this balancing out can only be so on average over all the charge carriers and there will be some deflection of charges moving with velocities higher and lower than the mean velocity.

It is usual to denote the magnitude of the Hall effect by the coefficient R_H where

$$R_H = V_H t / BI. \tag{6.15}$$

t is the thickness of the sample which in practice should be kept reasonably small in order that a large Hall voltage is established. Note that I is the total current through the sample, not the current density.

Alternatively, we can write

$$E_H = V_H / a = R_H BJ \tag{6.16}$$

where a is the width of the sample and J is the current density. Equation (6.16) can be compared with equation (3.8):

$$E_i = \rho_{ij} J_j. \tag{6.17}$$

Comparing equations (6.16) and (6.17), we can write the resistivity tensor in the form

$$\rho_{ij} = \rho_{ij}^0 + \rho_{ijk} B_k + \rho_{ijkl} B_k B_l. \tag{6.18}$$

We now have ρ_{ij}^0 as the resistivity tensor in the absence of a magnetic field. ρ_{ijk} is a third-order tensor representing the Hall effect which is a transverse effect assuming E and B are orthogonal. ρ_{ijkl} is a further higher-order effect representing a quadratic change in longitudinal resistance in a magnetic field. If B_k changes to $-B_k$ and B_l changes to $-B_l$, $B_k B_l$ remains unchanged. Hence, ρ_{ijkl} must be centrosymmetric whereas ρ_{ijk} is antisymmetric.

As we have seen already (p.53), it is possible to divide any second-rank tensor into a symmetrical and an antisymmetrical part. Thus we could write

$$E_i = (s_{ij} + a_{ij})j_j \qquad (6.19a)$$

where

$$s_{ij} = (\rho_{ij} + \rho_{ji})/2 = s_{ji} \qquad (6.19b)$$

and is the symmetrical part.

$$a_{ij} = (\rho_{ij} - \rho_{ji})/2 = -a_{ji} \qquad (6.19c)$$

and is the antisymmetrical part. Also, it can be seen that

$$s_{ij}(B) = s_{ij}(-B) \qquad (6.20a)$$

and

$$a_{ij}(B) = -a_{ij}(-B). \qquad (6.20b)$$

The symmetrical part contains only even functions of B. In effect, s_{ij} is a generalisation of magnetoresistance, and a_{ij} is a generalisation of the Hall effect. Assuming that we do not need to go to B^3 terms, then

$$a_{ij} = \rho_{ijk}B_k. \qquad (6.21)$$

We can see that the Hall coefficient relates the axial vector B_k to the antisymmetrical second-order tensor a_{ij} and in this form is a third-rank axial tensor. Representing the Hall effect in the form a_{ij} is representing it as a second-rank antisymmetrical tensor. Both approaches are used; they are equivalent because of equation (6.19c).

6.8 The Hall Effect; Relationship to Symmetry

Coefficients for the Hall effect are independent of direction for isotropic materials and for all cubic crystals. Otherwise, the number of independent coefficients varies from two for the more symmetric classes of the hexagonal, trigonal and tetragonal systems to nine for the triclinic system. For many of the properties considered so far, the tensor coefficients have fixed values for any single-crystal sample of a particular material. The values may vary slightly due to imperfections or impurities but they measure absolute properties of the crystal. The Hall effect is different. Its magnitude depends on the charge carrier concentration in the crystal, usually a semiconductor, and this concentration may vary considerably from sample to sample. In many isotropic samples, the Hall coefficient is given by

$$R_H = 1/ne$$

where n is the concentration of electrons and e is the electronic charge.

If R_H can take on a wide range of values for different samples of the same semiconductor but containing different amounts of impurity, one might question the value of measuring the coefficient. However, used in conjunction with simultaneously measured values of electrical conductivity, the values of R_H can tell one much about the electronic structure of the semiconductor. When measuring the Hall effect in a single crystal, it is particularly useful to know which components to measure and hence how to cut differently oriented samples from a single crystal.

Let us continue with the tetragonal example $\bar{4}$. Axes transform as before according to

$$1 \rightarrow -2 \quad 2 \rightarrow 1 \quad 3 \rightarrow -3.$$

Again when considering what happens to the tensor components we need to insert a minus sign overall as the direction of rotation along each new axis is reversed.

Considering B_1 first we have the subscript changing due to the symmetry operation including the rotational effect as

$$\begin{bmatrix} 111 & 121 & 131 \\ 211 & 221 & 231 \\ 311 & 321 & 331 \end{bmatrix} \rightarrow \begin{bmatrix} +222 & -212 & +232 \\ -122 & +112 & -132 \\ +322 & -312 & +332 \end{bmatrix}.$$

Taking B_2 next

$$\begin{bmatrix} 112 & 122 & 132 \\ 212 & 222 & 232 \\ 312 & 322 & 332 \end{bmatrix} \rightarrow \begin{bmatrix} -221 & +211 & -231 \\ +121 & +111 & +131 \\ -321 & +311 & -331 \end{bmatrix}.$$

And finally B_3

$$\begin{bmatrix} 113 & 123 & 133 \\ 213 & 223 & 233 \\ 313 & 323 & 333 \end{bmatrix} \rightarrow \begin{bmatrix} +223 & -213 & +233 \\ -123 & +113 & -133 \\ +323 & -313 & +333 \end{bmatrix}$$

$$111 = \quad 222 = -111 = 0 \qquad 131 = \quad 232$$

$$121 = -212 = -121 = 0 \qquad 231 = -132$$

$$211 = -122 = -211 = 0 \qquad 311 = \quad 322$$

$$221 = \quad 112 = -221 = 0 \qquad 321 = -312$$

$$331 = \quad 332 = -331 = 0 \qquad 113 = \quad 223$$

$$133 = \quad 233 = -133 = 0 \qquad 123 = -213$$

$$313 = \quad 323 = -313 = 0 \qquad 333 = \quad 333.$$

Also $\rho_{ijk} = -\rho_{jik}$ as interchanging electric current and magnetic field reverses the direction of the resultant Hall voltage; i.e. a_{ij} is antisymmetric. Hence,

$$131 = -311 = -322 = 232$$

$$132 = -312 = -231 = 321$$

$$113 = -113 = 0$$

$$333 = -333 = 0.$$

Consequently the resultant scheme of components for the Hall effect in a $\bar{4}$ crystal is

$$\begin{bmatrix} 0 & 0 & \rho_{131} \\ 0 & 0 & \rho_{132} \\ -\rho_{131} & -\rho_{132} & 0 \end{bmatrix}$$

$$\begin{bmatrix} 0 & 0 & \rho_{132} \\ 0 & 0 & \rho_{131} \\ -\rho_{132} & -\rho_{131} & 0 \end{bmatrix}$$

$$\begin{bmatrix} 0 & \rho_{123} & 0 \\ -\rho_{123} & 0 & 0 \\ 0 & 0 & 0 \end{bmatrix}.$$

Note that ρ_{131} and the equivalent components do not fit the classical Hall configuration with B orthogonal to E so that there are only two components of the type which are measured in the standard Hall apparatus.

6.9 Magnetoresistance and Other Effects

In equation (6.18) we had the quadratic term $\rho_{ijkl}B_kB_l$, the longitudinal magnetoresistance. Using symmetry arguments, the number of possible components is increased compared with a third-rank tensor. It becomes difficult to measure the coefficients independently and the subject becomes somewhat specialised.

In Chapter 3, we drew a particular parallel between electrical conductivity and thermal conductivity and there are parallel thermal effects in a magnetic field. The thermal conductivity of a crystal sample in the presence of a magnetic field can be expressed as

$$k_{ij} = k_{ij}^0 + k_{ijk}B_k + k_{ijkl}B_kB_l \qquad (6.22)$$

where k_{ij}^0 is thermal conductivity with zero magnetic field. The term involving k_{ijk} is called the Righi–Leduc effect and involves establishing a transverse temperature gradient when a longitudinal one is set up along the sample; not an especially useful effect. The term involving k_{ijkl} is the magnetothermal conductivity, which is sometimes called the Maggi–Righi–Leduc effect), and is the change in thermal conductivity in a magnetic field. Measurement of this effect in semiconductors can sometimes give useful data.

A full analysis of transport phenomena is beyond the scope of this book. However, having established the nature (polar or axial) of any tensor property and the appropriate crystal symmetry, working out the independent coefficients follows the procedures outlined.

Problems

6.1 Deduce the forms of the gyration tensor for orthorhombic crystals of class (a) 222 and (b) mm2. (In (a) the crystal exhibits twofold rotational symmetry along three orthogonal reference axes; in (*b*) there are two mirror planes perpendicular to two references axes and a twofold rotation axis parallel to the third reference axis, taken as x_3.)

6.2 Show that it is necessary to measure two components of the Hall effect in a cubic crystal of point group m3. (An m3 crystal has threefold rotational symmetry about its diagonals and a mirror plane perpendicular to the x_3 axis. The combination of these lead to twofold rotational axes along x_1, x_2 and x_3.)

Note: Two components are required for crystal classes m3 and 23 but only one component for the other two cubic classes and for isotropic media.

7 Optoelectronic Effects

7.1 Introduction

In many crystals applying an electric field alters the dielectric constant and the refractive index of the crystalline material. The applied electric field is in a particular direction and as a consequence an anaxial (or isotropic) material can become uniaxial and doubly refracting. Uniaxial and biaxial crystals will show changes in their refractive index components. According to the crystal class (see table 7.1 for examples of crystals in the different crystal classes), and hence the crystal coefficients, and according to the orientation of the crystal, the electric field may be applied longitudinally or transversely to the directon of passage of the light. In electro-optics, we are interested usually in both the linear and quadratic effects, so we write the change of optical impermeability in the form

$$\Delta \eta_{ij} = \eta_{ij}(\boldsymbol{E}) - \eta_{ij}(0)$$
$$= r_{ijk} E_k + s_{ijkl} E_k E_l. \tag{7.1}$$

The r_{ijk} constants are the linear electro-optic coefficients and are often called the Pockels coefficients as the linear effect is called the Pockels effect. The s_{ijkl} components are the quadratic or Kerr electro-optic coefficients†. The linear effect will not occur in centrosymmetric crystals; so although the quadratic effect can be expected to be small when both effects are present, it may be important in centrosymmetric crystals where the linear effect is

---

† s_{ijkl} is being used for Kerr coefficients in this edition, instead of p_{ijkl} as it has become common useage. The notation should not be confused with that for elastic stiffness.

absent. In general, the quadratic coefficients s_{ikjl} and also the linear coefficients r_{ijk} will depend on the wavelength of the light and the modulation frequency, as well as varying with temperature. Consequently, the electro-optic coefficients need to be known over a range of parameters for each crystalline material in use. We will consider the linear effect first. The quadratic effect will be considered later.

7.2 The Linear Electro-optic Effect (the Pockels Effect)

The optical indicatrix in the presence of the electric field E can be written as

$$\eta_{ij}(E)x_i x_j = 1 \qquad (7.2)$$

which reduces to

$$\eta_{ij}x_i x_j = 1$$

in the absence of an electric field E. The η tensor is symmetric and i and j can be permutated. We can write contracted subscripts for the linear electro-optic coefficients such that

$$r_{11k} = r_{1k} \qquad r_{22k} = r_{2k} \qquad r_{33k} = r_{3k}$$
$$r_{23k} = r_{32k} = r_{4k} \qquad r_{13k} = r_{31k} = r_{5k}$$
$$r_{12k} = r_{21k} = r_{6k} \qquad k = 1,\ 2,\ 3.$$

Using these contracted subscripts, the equation of the indicatrix in the presence of the electric field, assuming linear terms only, becomes

$$\left(1/n_1^2 + r_{1k}E_k\right)x_1^2 + \left(1/n_2^2 + r_{2k}E_k\right)x_2^2 + \left(1/n_3^2 + r_{3k}E_k\right)x_3^2$$
$$+ 2r_{4k}E_k x_2 x_3 + 2r_{5k}E_k x_3 x_1 + 2r_{6k}E_k x_1 x_2 = 1 \qquad (7.3)$$

where $k = 1, 2, 3$ and the E_ks are the components of the electric field in the x_1, x_2 and x_3 directions. The ellipsoid has to be written in the above form as in general the principal axes of this ellipsoid do not coincide with the principal axes x_1, x_2 and x_3 before application of the electric field.

Table 7.1 Examples of different crystals in different crystal systems and classes. (At least one example is included for each crystal class.)

Crystal	Class	Examples of crystals
Cubic (isotropic)	m3m	$BaTiO_3 (T > T_c)$, Ge, Si, $SrTiO_3$
	432	$LiFe_5O_8$
	$\bar{4}3m$	$Bi_4Ge_3O_{12}$, CdTe, GaAs, GaP, InAs, MgO, ZnSe, β-ZnS, ZnTe
	m3	MgO_2, $MnSe_2$
	23	$Bi_{12}Ge\,O_{20}$, $Bi_{12}SiO_{20}$
Hexagonal (uniaxial)	6/mmm	β-Al_2O_3, BN, NiAs
	622	$BaAl_2O_4$
	6mm	BeO, CdS, CdSe, ZnO, α-ZnS
	$\bar{6}$m2	GaS, GaSe
	6/m	β-Si_3N_4
	$\bar{6}$	LiO_2
	6	$LiIO_3$
Trigonal (uniaxial)	$\bar{3}$m	Bi_2Se_3, Bi_2Te_3, $CaCO_3$ (calcite)
	32	SiO_2 (α-quartz)
	3m	Ag_3AsS_3, Ag_3SbS_3, α-quartz, $LiNbO_3$, $LiTaO_3$
	$\bar{3}$	AsI_3, Ge_3M_4, S
	3	B_2O_3
Tetragonal (uniaxial)	4/mmm	GeO_2, TiO_2
	422	TeO_2
	$\bar{4}$2m	ADP, KDP
	4mm	$BaTiO_3$ $(T < T_c)$
	4/m	$AgIO_4$
	$\bar{4}$	$CdGa_2S_4$
	4	$Ba_2Nb_3O_{15}$, γ-Fe_2O_3
Orthorhombic (biaxial)	mmm	Al_2BeO_4 (Alexandrite), Bi_2S_3, $CaCl_2$
	22	α-HIO_3, Rochelle salt
	2mm	$KNbO_3$
Monoclinic (biaxial)	2/m	$AgAuTe_4$, Bi_2O_3, $PbSiO_3$
	2	KOH
	m	α-GaS_3, $PbHPO_4$, $SrTeO_3$
Triclinic (biaxial)	1	KIO_3
	$\bar{1}$	mica

7.3 The Pockels Effect in Lithium Niobate

Let us look at the linear electro-optic effect in lithium niobate, $LiNbO_3$, which has crystal symmetry 3m. That is, it is classed as trigonal and can be represented by a rhombohedral or a hexagonal cell. The rhombohedral cell contains two molecules whereas the hexagonal cell contains six and is the easier cell to use. Also, it is easier to use the hexagonal system of indices for directions and planes. The point group has its threefold rotational axis parallel to the [0001] direction and mirror planes of type $\{2\bar{1}\bar{1}0\}$ (figure 7.1). The electric field E is applied along the x_3 (i.e. the c) axis.

The forms of the components for the electro-optic tensor will depend on the definition of the mirror planes. Figure 7.1 shows one of the mirror planes perpendicular to x_1. (The other standard description is to define the mirror plane perpendicular to x_2.) The r_{ijk} coefficients must be symmetric in ij so we can reduce r_{ijk} to r_{mn} with $m = 1, \ldots, 6$, $n = 1, \ldots, 3$, following the pattern used previously. Carrying out the symmetry operations for point group 3m reduces the electro-optic coefficients to the form

$$\begin{pmatrix} 0 & -r_{22} & r_{13} \\ 0 & r_{22} & r_{13} \\ 0 & 0 & r_{33} \\ 0 & r_{51} & 0 \\ r_{51} & 0 & 0 \\ -r_{22} & 0 & 0 \end{pmatrix}.$$

We can now insert the coefficients into equation (7.3) for the indicatrix. $k = 3$ with E along x_3. The cross-product terms in x_1, x_2 and x_3 disappear as r_{43}, r_{53} and r_{63} are zero. The principal axes remain the x_1, x_2 and x_3 axes after the electric field is applied.

$$\left(1/n_O^2 + r_{13}E\right)x_1^2 + \left(1/n_O^2 = r_{13}E\right)x_2^2 + \left(1/n_E^2 + r_{33}E\right)x_3^2 = 1. \quad (7.4)$$

Here we have put n_1 and n_2 equal to n_O, the ordinary refractive index before application of the electric field. Similarly, n_3 has been put equal to n_E, the extraordinary refractive index, also before application of the electric field.

The lengths of the semiaxes become, after application of the electric field E,

$$n_{x_1} = n_O - n_O^3 r_{13}E/2$$

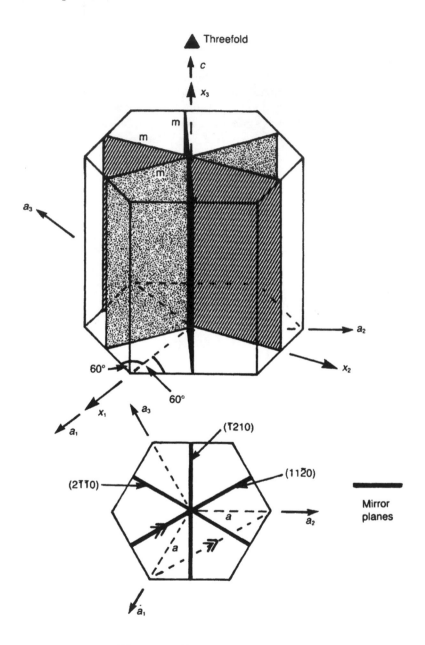

Figure 7.1 The 3m hexagonal cell.

$$n_{x_2} = n_O - n_O^3 r_{13} E / 2$$

and

$$n_{x_3} = n_E - n_E^3 r_{33} E / 2.$$

For a light beam passing along the x_1 axis, the birefringence becomes

$$n_{x_3} - n_{x_2} = \left(n_E - n_O \right) - \left(n_E^3 r_{33} - n_O^3 r_{13} \right) E / 2. \qquad (7.5)$$

We can show diagrammatically (figure 7.2) the effect on the median section of the refractive index ellipsoid of applying the electric field parallel to the optic axis x_3 (i.e. axis c). We can write the change of refractive index for light in the x_1 direction as

$$\Delta n = \Delta n_{E=0} + \alpha E \qquad (7.6)$$

where

$$\alpha = \frac{1}{2} \left(n_O^3 r_{13} - n_E^3 r_{33} \right).$$

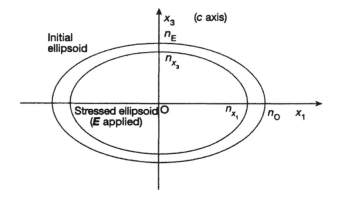

Figure 7.2 Refractive index ellipsoids for lithium niobate in the Ox_1x_3 plane with no applied electric field and with electric field applied in the Ox_3 direction.

When we insert values for n_O, n_E, r_{13} and r_{33} (see table 7.2), the magnitude of α is 1.1×10^{-10} m V^{-1}. Examples of magnitudes of certain optoelectronic effects in selected materials are calculated later (pp.126–9).

Thus by altering the magnitude of the electric field, and hence the birefringence, any desired polarisation state can be obtained for a particular thickness of crystal. In addition, the polarisation state can be modulated using a sinusoidal applied electric field. Using crossed linear polarisers at the input and output stages, the light can be amplitude modulated, with zero output when no electric field is applied, and maximum output when the electric field is of the correct magnitude to rotate the polarised beam by 90°.

7.4   The Pockels Effect in ADP (Ammonium Dihydrogen Phosphate)

ADP belongs to the class $4\bar{2}m$ and, with a higher symmetry than that of lithium niobate, there are electro-optic coefficients in the matrix that become zero. Similar arguments apply to KDP (potassium dihydrogen phosphate) which has the same crystal class. Again carrying out the symmetry operations, the required matrix now becomes

$$
\begin{pmatrix}
0 & 0 & 0 \\
0 & 0 & 0 \\
0 & 0 & 0 \\
r_{41} & 0 & 0 \\
0 & r_{41} & 0 \\
0 & 0 & r_{63}
\end{pmatrix}.
$$

Although both ADP and KDP are uniaxial materials, the presence of the r_{63} term means that applying an electric field produces a crossed term in establishing the modified indicatrix ellipsoid. We will consider the two special cases of an electric field parallel and perpendicular to the optic axis.

For the electric field parallel to x_3 (the c axis), substituting into equation (7.3) $E_k = E_3 = E$, also $n_1 = n_2 = n_O$, and $n_3 = n_E$, we obtain

$$
\frac{x_1^2 + x_2^2}{n_O^2} + \frac{x_3^2}{n_E^2} + 2r_{63}Ex_1x_2 = 1. \tag{7.7}
$$

The modified ellipsoid is no longer symmetrical around the x_3 axis and the material has become biaxial. The circular cross section of the indicatrix ellipsoid becomes elliptical and has new axes Ox_1' and Ox_2' rotated by $\pi/4$ from the original axes Ox_1 and Ox_2 respectively. The new semiaxes have changed to

$$n_{x_1'} = n_O - \tfrac{1}{2}n_O^3 r_{63}E \quad \text{and} \quad n_{x_2'} = n_O + \tfrac{1}{2}n_O^3 r_{63}E \quad (7.8)$$

(see figure 7.3). Hence a slice of crystal cut perpendicular to the Ox_3 axis will act like a birefringent plate with a fast axis Ox_1' and a slow axis Ox_2' with a birefringence

$$\Delta n = n_{x_1'} - n_{x_2'} = n_O^3 r_{63}E. \quad (7.9)$$

This configuration can be used for optical switching (see p.120).

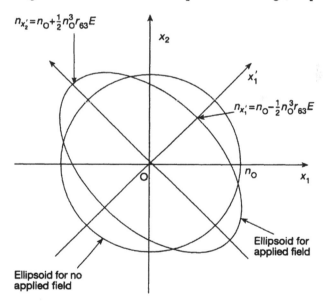

Figure 7.3 Refractive index ellipsoids for ADP in the $Ox_1 x_2$ plane with no applied electric field and with electric field applied in the Ox_3 direction.

For an electric field in the $Ox_1 x_2$ plane with the electric field making an angle α with the Ox_1 axis, equation (7.3) becomes

$$\frac{x_1^2 + x_2^2}{n_O^2} + \frac{x_3^2}{n_E^2} + 2r_{41}E(\cos\alpha x_2 x_3 + \sin\alpha x_1 x_3) = 1. \quad (7.10)$$

The only configurations which give a simple mathematical result are for $\alpha = 0$ (the electric field is in the Ox_1 direction) and $\pi/2$ (the electric field is in the Ox_2 direction). The results are the same for each. Even in these two cases we cannot match up the two axes of the ellipsoid simply with the null field situation (note that x_3 is included in a cross-product term in the equation).

So using $\alpha = \pi/2$ (electric field in the Ox_2 drection), we have

$$\frac{x_1^2 + x_2^2}{n_O^2} + \frac{x_3^2}{n_E^2} + 2r_{41}E\alpha x_1 x_3 = 1. \tag{7.11}$$

As the cross-product term $x_1 x_3$ does not involve x_2, the axis Ox_2 remains an axis of the new ellipsoid. It can be shown that a rotation β in the $x_1 x_3$ plane given by

$$\tan\beta = \frac{2r_{41}En_O^2 n_E^2}{n_E^2 - n_O^2}$$

gives an ellipsoid of the form

$$\frac{x_{1'}^2}{n_{x_1'}^2} + \frac{x_{2'}^2}{n_O^2} + \frac{x_{3'}^2}{n_{x_3'}^2} = 1$$

where

$$n_{x_1'} = n_O - \tfrac{1}{2}n_O^3 r_{41}E\tan\beta \quad \text{and} \quad n_{x_3'} = n_E - \tfrac{1}{2}n_E^3 r_{41}E\tan\beta. \tag{7.12}$$

The result is summarised in figure 7.4 where we now have a cross section of the ellipsoid in the $x_1 x_3$ plane. As β is small, $\tan\beta \approx \beta$, and $\tan\beta$ is proportional to E. Hence the birefringence takes the form

$$\Delta n = n_{x_3'} - n_{x_1'} = \Delta n_{E=0} + \alpha E^2 \tag{7.13}$$

and we see a modification to the birefringence which is now proportional to the square of the applied electric field.

7.5   The Pockels Effect in Gallium Arsenide

Gallium arsenide (GaAs) has crystal class $\bar{4}3m$, as does indium arsenide and cadmium telluride (see table 7.1). This is a high

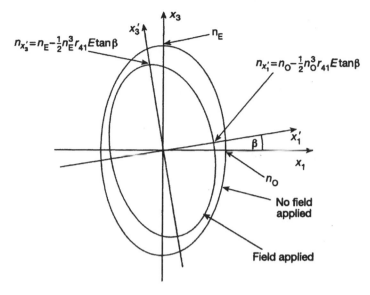

$n_{x_3'} = n_E - \frac{1}{2}n_E^3 r_{41} E \tan\beta$

$n_{x_1'} = n_O - \frac{1}{2}n_O^3 r_{41} E \tan\beta$

Figure 7.4 Refractive index ellipsoids for ADP in the $Ox_1 x_3$ plane with no applied electric field and with an electric field in the Ox_2 direction.

symmetry class and there is only one independent term

$$\begin{pmatrix} 0 & 0 & 0 \\ 0 & 0 & 0 \\ 0 & 0 & 0 \\ r_{41} & 0 & 0 \\ 0 & r_{41} & 0 \\ 0 & 0 & r_{41} \end{pmatrix}$$

so we can use $r = r_{41}$ and if we use n for the isotropic refractive index before the electric field is applied, equation (7.3) gives

$$\frac{1}{n^2}\left(x_1^2 + x_1^2 + x_1^2\right) + 2r\left(x_2 x_3 E_1 + x_1 x_3 E_2 + x_1 x_2 E_3\right) = 1. \quad (7.14)$$

The interesting cases are with the electric field in the $\langle 001 \rangle$, $\langle 110 \rangle$ and $\langle 111 \rangle$ directions. We now show the case for the $\langle 110 \rangle$ direction, whereas the $\langle 111 \rangle$ direction is illustrated by question 7.3 at the end of the chapter.

For the $\langle 110 \rangle$ case, the electric field E is given by $\frac{1}{\sqrt{2}}(\hat{x}_1 + \hat{x}_2)$ where $\hat{x}_1$ and $\hat{x}_2$ are unit vectors. For the $\langle 110 \rangle$ case the ellipsoid can be written in the form

$$\frac{1}{n^2}\left(x_1^2 + x_1^2 + x_1^2\right) + 2rE\left(x_2x_3 + x_1x_3\right) = 1 \qquad (7.15)$$

and the modified refractive indices become (for $rE \ll 1$)

$$n_{x_1'} = n + \frac{1}{2}n^3rE \qquad n_{x_2'} = n - \frac{1}{2}n^3rE \qquad n_{x_3'} = n \qquad (7.16)$$

corresponding to axes (figure 7.5(a))

$$\hat{x}_1' = \frac{1}{2}\left(\hat{x}_1 + \hat{x}_2 - \sqrt{2}\hat{x}_3\right)$$

$$\hat{x}_2' = \frac{1}{2}\left(\hat{x}_1 + \hat{x}_2 + \sqrt{2}\hat{x}_3\right)$$

$$\hat{x}_3' = \frac{1}{\sqrt{2}}\left(\hat{x}_1 - \hat{x}_2\right).$$

That is the axes have rotated relative to all three original axes and the situation is not easy to visualise. However, the modified refractive index ellipsoid follows a similar pattern to earlier.

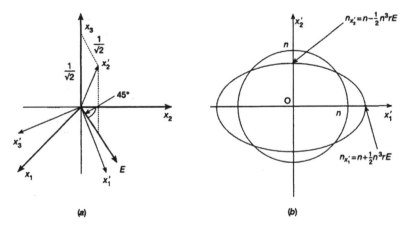

(a)   (b)

Figure 7.5 (a) Modified axes for gallium arsenide when an electric field is applied in the $\langle 110 \rangle$ direction and (b) the refractive index ellipsoid sections in the $Ox_1'x_2'$ plane.

For plane wave propagation in the x_3' direction perpendicular to the plane containing axis Ox_3 and electric field E (figure 7.5(b)), the birefringence is

$$\Delta n = n_{x_1'} - n_{x_2'} = n^3 r E. \tag{7.17}$$

This configuration is used for optical modulators in a transverse arrangement.

7.6   The Quadratic Electro-optic Effect (The Kerr Effect)

This can occur in crystals of any symmetry. The pairs of subscripts i, j and k, l of the tensor components s_{ijkl} permute. However, there are no arguments as to why s_{mn} in the reduced subscript notation should be the same as s_{nm}. So there are a maximum of 36 components before crystal symmetry criteria are incorporated compared with the 21 for elasticity.

We see from equation (7.1) that we need to replace terms of the form $r_{ijk}E_k$ in the linear effect with terms of the form $s_{ijkl}E_kE_l$ to obtain the quadratic effect. We continue to use the contracted notation for subscript pairs: we use the notations here for both pairs ij and kl. As a consequence we can see that the first term in equation (7.2) involving x_1^2 becomes in the quadratic electro-optic case

$$\left(1/n_1^2 + s_{11}E_1^2 + s_{12}E_2^2 + s_{13}E_3^2 + 2s_{14}E_2E_3 \right.$$
$$\left. + 2s_{15}E_3E_1 + 2s_{16}E_1E_2 \right)x_1^2.$$

Coefficients are obtained similarly for terms in x_2^2 and x_3^2.

The term involving x_2x_3 becomes

$$2\left(s_{41}E_1^2 + s_{42}E_2^2 + s_{43}E_3^2 + 2s_{44}E_2E_3 \right.$$
$$\left. + 2s_{45}E_3E_1 + 2s_{46}E_1E_2 \right)x_2x_3$$

and we have similar coefficients for x_3x_1 and x_1x_2.

7.7   The Kerr Effect in Barium Titanate

A material which can be used as a Kerr device is barium titanate, $BaTiO_3$. Below $120\,°C$ the crystal has tetragonal structure of class 4mm; it is not centrosymmetric and the linear effect dominates. Above $120\,°C$ the crystal changes phase to become cubic of class m3m when it is centrosymmetric. So above the $120\,°C$ transition temperature, the linear effect disappears.

A suitable direction in which to apply the electric field is any one of the $\langle 110 \rangle$ directions. Assuming the particular direction [110], the electric field E can be resolved into components of magnitude $E/\sqrt{2}$ along x_1 and along x_2; i.e. $E_1 = E_2 = E/\sqrt{2}$ and $E_3 = 0$. By applying symmetry arguments, the pattern of components can be shown to be similar to that of compliance in class 4 (p.63), i.e.

$$\begin{pmatrix} s_{11} & s_{12} & s_{12} & 0 & 0 & 0 \\ s_{12} & s_{11} & s_{12} & 0 & 0 & 0 \\ s_{12} & s_{12} & s_{11} & 0 & 0 & 0 \\ 0 & 0 & 0 & s_{44} & 0 & 0 \\ 0 & 0 & 0 & 0 & s_{44} & 0 \\ 0 & 0 & 0 & 0 & 0 & s_{44} \end{pmatrix}.$$

The indicatrix takes the form

$$\left(1/n^2 + s_{11}E^2/2 + s_{12}E^2/2\right)x_1^2 + \left(1/n^2 + s_{11}E^2/2 \right.$$
$$\left. + s_{12}E^2/2\right)x_2^2 + \left(1/n^2 + s_{12}E^2\right)x_3^2 + 2x_1x_2s_{44}E^2 = 1.$$

$$(7.18)$$

All the other terms are zero. For instance, all terms involving x_2x_3 and x_3x_1 are zero. The only non-zero term involving x_1x_2 has the coefficient $2s_{66}E_1E_2$ which equals $2s_{44}E^2$.

To get the ellipsoid into a form related to the principal axes in the presence of the applied electric field, it is necessary to rotate the axes by $45°$ in the x_1x_2 plane—i.e. to give a principal axis along the direction of the electric field, which becomes a direction of symmetry.

We can draw up a direction cosine table for the cosines of the angle between the old and the new (primed) axes:

	x_1'	x_2'	x_3'	Σl^2
x_1	$1/\sqrt{2}$	$-1/\sqrt{2}$	0	1
x_2	$1/\sqrt{2}$	$1/\sqrt{2}$	0	1
x_3	0	0	1	1
Σl^2	1	1	1	

To change to the new system of axes we obtain expressions for the old coordinates in terms of the new coordinates.

$$x_1^2 = \left(x_1'^2 + x_2'^2 - 2x_1'x_2'\right)/2$$

$$x_2^2 = \left(x_1'^2 + x_2'^2 + 2x_1'x_2'\right)/2$$

$$x_1^2 + x_2^2 = x_1'^2 + x_2'^2$$

$$x_1 x_2 = \left(x_1'/\sqrt{2} - x_2'/\sqrt{2}\right)\left(x_1'/\sqrt{2} + x_2'/\sqrt{2}\right)$$

$$= x_1'^2/2 - x_2'^2/2$$

$$x_3^2 = x_3'^2.$$

Equation (7.18) becomes

$$\left(1/n^2 + s_{11}E^2/2 + s_{12}E^2/2 + s_{44}E^2\right)x_1^2 + \left(1/n^2 + s_{11}E^2/2\right.$$
$$\left. + s_{12}E^2/2 - s_{44}E^2\right)x_2^2 + \left(1/n^2 + s_{12}E^2\right)x_3^2 = 1. \quad (7.19)$$

So comparing this equation with the standard equation for the indicatrix, equation (5.3), we have

$$\left(n_{x_1}\right)^{-2} = n^{-2} + \left(s_{11} + s_{12}\right)E^2/2 + s_{44}E^2$$

$$\left(n_{x_2}\right)^{-2} = n^{-2} + \left(s_{11} + s_{12}\right)E^2/2 - s_{44}E^2$$

$$\left(n_{x_3}\right)^{-2} = n^{-2} + s_{12}E^2.$$

Assuming that the overall changes to n are small, we can expand to third order in n to obtain

$$n_{x_1} = n - n^3\left(s_{11} + s_{12}\right)E^2/4 - n^3 s_{44}E^2/2$$

$$n_{x_2} = n - n^3\left(s_{11} + s_{12}\right)E^2/4 + n^3 s_{44}E^2/2$$

$$n_{x_3} = n - n^3 s_{12}E^2/2. \quad (7.20)$$

7.8 The Kerr Effect in ADP and KDP

As already indicated (p.112), these materials belong to crystal class $4\bar{2}m$ and the matrix for the Kerr effect can be written as

$$
\begin{pmatrix}
s_{11} & s_{12} & s_{13} & 0 & 0 & 0 \\
s_{12} & s_{11} & s_{13} & 0 & 0 & 0 \\
s_{31} & s_{31} & s_{33} & 0 & 0 & 0 \\
0 & 0 & 0 & s_{44} & 0 & 0 \\
0 & 0 & 0 & 0 & s_{44} & 0 \\
0 & 0 & 0 & 0 & 0 & s_{66}
\end{pmatrix}
$$

We will apply the electric field in the Ox_3 direction. The equation for the indicatrix becomes (compare with equation (7.18) for $BaTiO_3$):

$$
\left(\frac{1}{n_O^2} + s_{13}E^2\right)x_1^2 + \left(\frac{1}{n_O^2} + s_{13}E^2\right)x_2^2 + \left(\frac{1}{n_E^2} + s_{13}E^2\right)x_3^2 = 1. \quad (7.21)
$$

The crystalline material remains uniaxial with optic axis Ox_3 but the birefringence alters from $\Delta n_{E=0}$ to $\Delta n'$ where

$$
\Delta n' = \Delta n_{E=0} - \frac{1}{2}\left(n_E^3 s_{33} - n_O^3 s_{13}\right)E^2. \quad (7.22)
$$

From equation (7.8) on p.113, the first-order change in birefringence (i.e. the change due to the Pockels effect) was $\frac{1}{2}n_O^3 r_{63}E$. Here the second-order change due to the Kerr effect is $\frac{1}{2}(n_O^3 s_{33} - n_E^3 s_{13})E^2$. The ratio of the size of the second-order effect to the first-order effect (assuming that $n_E \approx n_O$) is

$$
\frac{s_{33} - s_{13}}{r_{63}} E.
$$

7.9 Optical Switching, Intensity Modulation and Phase Modulation

A typical device for switching using the Pockels effect can be cut from ADP or KDP crystals with faces perpendicular to the optical axis Ox_3 (i.e. perpendicular to the c axis) (figure 7.6). For a monochromatic beam of light travelling in the x_3 direction, the crystal appears to be isotropic. Applying an electric field parallel to the x_3 axis produces birefringence Δn of magnitude $n_O^3 r_{63} E$, that is equal to $n_O^3 r_{63} \frac{V}{L}$ (see p.113), where V is the voltage applied between the end faces (note that transparent electrodes must be deposited on these end faces) and L is the length of the crystal. For light of wavelength λ, the birefrigence produces a phase difference δ of $(2\pi/\lambda)L\Delta n$ or

$$\delta = \frac{2\pi}{\lambda} n_O^3 r_{63} V \qquad (7.23)$$

and the birefringence can be varied by varying V. According to whether the analyser is parallel or perpendicular to the polariser, the observed intensity will be

$$I = \frac{1}{2} I_0 \cos^2 \frac{\phi}{2} \qquad \text{(parallel)}$$

or

$$I = \frac{1}{2} I_0 \sin^2 \frac{\phi}{2} \qquad \text{(perpendicular)}.$$

To act as a switch the intensity I needs to vary between zero and I_0. For crossed polarisers I will be zero for $V = 0$ and will equal I_0 for a value of V of $V_\pi = \lambda/2n_O^3 r_{63}$. When the voltage is V_π, the plate is acting as a half-wave plate; hence the notation.

A small intensity modulation can be superimposed on the intensity if required. It is best to superimpose this modulation onto a voltage of $V_\pi/2$, i.e. on $V_{\pi/2}$; that is the voltage is varied as

$$V = V_0 \sin \omega t + V_{\pi/2}$$

so

$$\phi = \frac{2\pi}{\lambda} n_O^3 r_{63} V_0 \sin \omega t + \pi/2.$$

With crossed polarisers

$$I = \frac{1}{2} I_0 \sin^2 \frac{\phi}{2} = \frac{1}{4} I_0 (1 - \cos \phi)$$

$$= \frac{1}{4} I_0 \left(1 + \sin \left(\frac{2\pi}{\lambda} n_O^3 r_{63} V_0 \sin \omega t \right) \right).$$

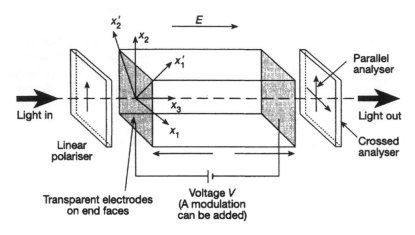

Figure 7.6 Electro-optic device (longitudinal configuration) using the Pockels effect.

Using $\sin \theta \approx \theta$ we have

$$I = \frac{1}{4} I_0 \left(1 + \frac{2\pi}{\lambda} n_O^3 r_{63} V_0 \sin \omega t \right). \qquad (7.24)$$

For phase modulation, a polariser is used in the correct orientation to allow light waves to pass parallel to one of the birefringent axes Ox_1' or Ox_2' of say ADP (see figure 7.3). Under this condition, application of an electric field does not change the state of the polarisation but it does change the output phase. The change of phase equals

$$\frac{\omega}{c} \Delta n_x' = \frac{n_O^3 r_{63} E_{x_3} L}{2c}.$$

Now the applied voltage V can be made sinusoidal such that

$$V = V_m \sin \omega_m t$$

where V_m is the amplitude of modulation and ω_m is the angular frequency. The output intensity I becomes proportional to

$$\cos^2 \left[\omega t - \frac{\omega}{c} n_O - \frac{n_O^3}{2} r_{63} V_m \sin \omega_m t \right]$$

which has the form

$$I = I_0 \cos^2\left[\omega t + \delta \sin \omega_m t\right] \qquad (7.25)$$

where $\delta = \pi n_O^3 r_{63} V_m / \lambda$ is called the phase modulation index. A correct analysis of the energy distribution requires the use of Bessel functions. It is interesting to note that δ is half the value of δ given by equation (7.23) for phase difference by birefringence.

7.10 Optical Beam Deflectors

Continuing with the use of ADP (or KDP) crystals, we can construct two half prisms with their edges parallel to the optic axis Ox_3'. The two prisms are placed on top of each other (figure 7.7) so that their Ox_3' axes are in opposite directions. Hence an electric field parallel to the x_3' direction acts as positive for one prism and negative for the other.

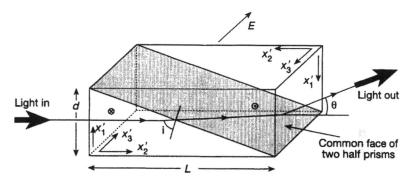

Figure 7.7 Optical beam deflector consisting of two ADP half-prisms.

In the lower prism (LP) we have refractive index given by

$$n_{LP} = n_O - \frac{n_O^3 r_{63} E_{x3}}{2}$$

and in the upper prism (UP) refractive index given by

$$n_{UP} = n_O + \frac{n_O^3 r_{63} E_{x3}}{2}.$$

By Snell's law

$$n_{LP} \sin i = n_{UP} \sin(i + \delta i)$$

where i is the angle of incidence at the common face of the prisms.

Now $\tan i = d/L$, so that $\delta i = \delta n d /(nL)$ where $\delta n = n_O^3 r_{63} E_{x_3}$. With refraction on the exit face, the angle of deflection θ becomes $n_O \delta i$ or

$$\theta = \frac{L}{d} n_O^3 r_{63} E_{x_3} \tag{7.26}$$

where E_{x_3} is the applied field which can be varied to vary the deflection angle of the light beam.

7.11   Non-linear Optics and Phase Matching

On p.106 we identified the Pockels effect as first order and the Kerr effect as second order within the equation

$$\Delta \eta_{ij} = \eta_{ij}(E) - \eta_{ij}(0) = r_{ijk} E_k + s_{ijkl} E_k E_l.$$

Alternatively we can express the difference as

$$\Delta \eta_{ij} = \eta_{ij}(P) - \eta_{ij}(0) = f_{ijk} P_k + g_{ijkl} P_k P_l \tag{7.27}$$

where P is the polarisation vector and f_{ijk} and g_{hjkl} are Pockels and Kerr coefficients defined using polarisation rather than electric fields. In non-linear optics it is customary to use polarisation P and write the polarisation in the form

$$P_i = \varepsilon_0 E_j + 2 d_{ijk} E_j E_k + 4 \chi_{ijkl} E_j E_k E_l \tag{7.28}$$

where χ_{ij}. d_{ijk} and χ_{ijkl} are the linear, second-order non-linear and third-order non-linear susceptibilities respectively. The optical responses given by various alternatives of the d_{ijk} and χ_{ijkl} give rise to a number of different non-linear effects not so far discussed including second-harmonic generation (frequency doubling) and third-order effects such as Raman and Brillouin scattering. Only the first of these effects will be discussed here.

A consequence of the above is that one can get non-linear coupling of two optical fields (using two light rays) and if one does the appropriate summation (involving $E_j^{\omega_1} e^{i\omega_1 t}$ and $E_k^{\omega_2} e^{i\omega_2 t}$ to represent the sinusoidal variation of the electric fields $E_j^{\omega_1}$ and $E_k^{\omega_2}$ at frequencies ω_1 and ω_2 respectively) one obtains

$$P_i^{\omega_3} = \omega_1 + \omega_2 = 2d_{ijk} E_j^{\omega_1} E_k^{\omega_2}. \tag{7.29}$$

If $\omega_1 = \omega_2$ and $\omega_3 = 2\omega_1$, the equation is often written in the contracted form

$$P_i^{2\omega} = 2d_{ijk} E_j^{\omega} E_k^{\omega} \tag{7.30}$$

where, as there is no physical significant difference between the jk and the kj components, the jk tensor notation suffixes are replaced by matrix notation suffixes 1 to 6 as shown on p.58. The resulting matrix components d_{mn} obey the same symmetry restrictions as the electro-optic matrix components r_{mn} (p.109). Magnitudes are different but related†. Hence there is a close connection between non-linear optics and optoelectronic effects.

For efficient second-harmonic generation it is not sufficient to merely combine the two beams; what is also required is that the second-harmonic waves generated at different planes within the crystal are all in phase (phase-matched) and this requires the wave-vector condition either that the wave vector for the frequency doubled wave is twice that for the incoming fundamental waves, $k^{2\omega} = 2k^{\omega}$, or alternatively $\Delta k = 0$, $k^{2\omega} - k^{\omega} = 0$. This can be achieved by matching the indices of refraction, $n^{2\omega} = n^{\omega}$, but the refractive index usually varies with frequency. Matching can be achieved in birefringent crystals by getting a match between the refractive indices for the ordinary and extraordinary rays. This approach can be seen clearly in figure 7.8 where the difference between the refractive index surfaces has been highly exaggerated. Surfaces are shown for the ordinary and extraordinary surfaces for angular frequencies ω and 2ω and it is easy to see that there is a direction Θ (measured relative to the x_3 direction, the optic axis) where Δk is zero. The example shown is for a negative uniaxial crystal and the condition for Θ is given by the equation

$$\frac{\cos^2 \Theta}{(n_O^{2\omega})^2} + \frac{\sin^2 \Theta}{(n_E^{2\omega})^2} = \frac{1}{(n_O^{\omega})^2}. \tag{7.31}$$

† It can be shown that they are related by $d_{ijk} = (-\epsilon_{ii}\epsilon_{jj}/4\epsilon_0)r_{ijk}$ and the quadratic nonlinear susceptibility is related to the Kerr coefficient by $\chi_{ijkl} = (-\epsilon_{ii}\epsilon_{jj}/12\epsilon_0)s_{ijkl}$.

We cannot here go into full details regarding the application of optoelectronics and non-linear optics, nor give the full details of the tensor components for the various effects in different materials. The reader is referred to a number of texts which give further information, including Yariv and Yeh (1984), Butcher and Cotter (1990) and Popov *et al* (1995).

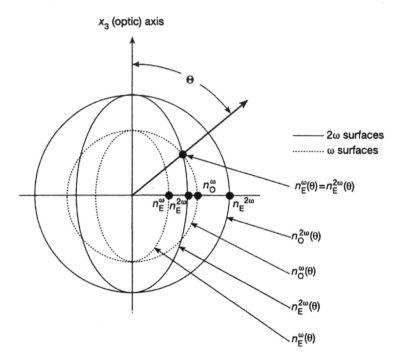

Figure 7.8 Phase matching between ordinary and extraordinary rays in a negative uniaxial crystal.

7.12   Worked Examples to Show Magnitudes in Actual Materials

To produce effects of sufficient size to be useful, large electric fields are usually required, particularly to establish second-order effects. However, suitable electric fields are produced relatively

Table 7.2 Examples of linear electro-optic coefficients and refractive indices. Coefficients are for a wavelength of 633 nm except where marked † when it is 10.6 μm.

Material	Symmetry	Electro-optic coefficients (pm V^{-1})	Refractive index
ADP	$\bar{4}2m$	$r_{41} = 23.4, r_{63} = -7.8$	$n_O = 1.5220,$ $n_E = 1.4773$
CdSe	6mm	$r_{33} = 4.3$	$n_O = 2.437,$ $n_E = 2.363$
CdTe†	$\bar{4}3m$	$r_{41} = 6.8$	$n = 2.60$
GaAs†	$\bar{4}3m$	$r_{41} = 1.51$	$n = 3.3$
GaP	$\bar{4}3m$	$r_{41} = -1.0$	$n = 3.6$
KDP	$\bar{4}2m$	$r_{41} = 8.6, r_{63} = -10.5$	$n_O = 1.5074,$ $n_E = 1.4669$
KNbO$_3$	2mm	$r_{13} = 28, r_{23} = 1.3,$ $r_{33} = 64, r_{42} = 380,$ $r_{51} = 105$	$n_1 = 2.280,$ $n_2 = 2.329,$ $n_3 = 2.169$
LiIO$_3$	6	$r_{13} = 4.1, r_{33} = 6.4,$ $r_{41} = 1.4, r_{51} = 3.3$	$n_O = 1.883,$ $n_E = 1.7367$
LiNbO$_3$	3m	$r_{13} - 9.6, r_{22} = 6.8,$ $r_{33} = 30.9, r_{51} = 32.6$	$n_O = 2.286,$ $n_E = 2.200$
LiTaO$_3$	3m	$r_{13} - 8.4, r_{22} = -0.2,$ $r_{33} = 30.5$	$n_O = 2.176,$ $n_E = 2.180$
ZnO	6mm	$r_{14} = 1.4, r_{33} = 2.6$	$n_O = 1.990,$ $n_E = 2.006$

easily in devices used in integrated optics. Tables 7.2 and 7.3 give selected data for commonly used material. Not only are there a large number of materials exhibiting electro-optic effects and each material has a number of relevant coefficients, but also the coefficients are usually dependent on wavelength. Consequently specific data are required for the actual operating conditions of the device in use.

Question 7A
What is the change of refractive index for a crystal of germanium

Table 7.3 Examples of quadratic electro-optic coefficients (wavelength 540 nm).

Material	Symmetry	Electro-optic coefficients $(10^{-18} \text{ m}^2\text{V}^{-1})$	Refractive index
ADP	$\bar{4}2m$	$n_E^2(s_{33} - s_{13}) = 24,$ $n_O^3(s_{31} - s_{11}) = 16.5,$ $n_O^3(s_{33} - s_{13}) = 5.8, n_O^3 s_{66} = 2$	$n_O = 1.5266,$ $n_E = 1.481$
KDP	$\bar{4}2m$	$n_E^3(s_{33} - s_{13}) = 31,$ $n_O^3(s_{31} - s_{11}) = 13.5,$ $n_O^3(s_{33} - s_{13}) = 8.9, n_O^3 s_{66} = 3$	$n_O = 1.512,$ $n_E = 1.4706$
PLZT†	∞m	$n^3(s_{33} - s_{13}) = 2600 \ (63\,^\circ\text{C})$	$n = 2.45$

† $Pb_{0.88}La_{0.8}Ti_{0.35}Zr_{0.65}O_3$.

when a voltage of 100 V is applied in the (001) direction to (i) a crystal of length 1 cm and (ii) a thin crystalline layer of thickness 100 μm for wavelength 10.6 μm and what is the phase retardation in each case?

Answer

(i) $\Delta n = n^3 r E = \dfrac{3.33 \times 1.51 \times 10^{-12} \times 100}{10^{-2}} = 5.4 \times 10^{-7}$

phase retardation $= \dfrac{2\pi}{\lambda} n^3 r E l = \dfrac{2\pi \times 5.4 \times 10^{-7}}{10.6 \times 10^{-6}} = 0.1\pi.$

(ii) $\Delta n = \dfrac{3.33 \times 1.51 \times 10^{-12} \times 100}{10^{-4}} = 5.4 \times 10^{-5}$

phase retardation $= 10\pi.$

This example demonstrates the small change of refractive index but the large corresponding phase retardation.

Question 7B

An optical beam deflector consisting of two ADP prisms as illustrated in figure 7.7 is used to deflect a laser beam of wavelength 633 nm. Calculate the magnitude of the deflection for a voltage of 1000 V applied to a crystal of length L of 1.0 cm and height d of 0.30 cm.

Answer

$$\theta = \frac{L}{d}n_O^3 r_{63}\frac{V}{L} = \frac{1.522^3 \times 7.8 \times 10^{-12} \times 1000}{0.3} = 0.92 \times 10^7 \text{ rad}$$
$$= 0.5 \times 10^{-3} \text{ deg.}$$

This would produce a deflection of 0.5 mm at a distance of 1 m.

Question 7C

Compare the size of the first-order Pockels change in birefringence (at 633 mm) to the second-order Kerr change of refractive index (at 540 nm) for KDP assuming a voltage of 150 V across a crystal thickness of 3.0 mm using data from tables 7.2 and 7.3.

Answer

$$\text{Ratio} \approx \frac{s_{33} - s_{13}}{r_{63}} E = \frac{31 \times 150 \times 1 \times 10^{-6}}{1.47^3 \times 3 \times 10^{-3} \times 7.8} = 0.02.$$

If we do not assume $n_O \approx n_E$, the ratio is

$$\sim \frac{n_E^3 \times 0.21}{n_O^3} = \left(\frac{1.48}{1.522}\right)^3 \times 0.021 \approx 0.019.$$

Problems

7.1 Show that when an electric field is applied along the c axis (i.e. in the [001] direction) of a class $\bar{4}$ crystal the indicatrix takes the form

$$\left(1/n_O^2 + r_{13}E\right)\left(x_1^2 + x_2^2\right) + x_3^2/n_E^2 + 2x_1x_2r_{63}E = 1.$$

The new principal axes can be obtained by clockwise rotation of the original axes (prior to application of the electric field) by angle θ looking along the positive x_3 direction. Show that

$$\theta = \frac{1}{2}\tan^{-1}\left(r_{63}/r_{13}\right).$$

Note that this angle is independent of E. Obtain expressions for the lengths of the principal axes of refraction. Hence show that for light propagating along the x_3 axis the birefringence is given by

$$n_O^3 E \left(r_{13}^2 - r_{63}^2 \right)^{1/2}.$$

7.2 Cadmium selenide has symmetry 6mm i.e. it has a sixfold rotational axis along the c direction and six mirror planes intersecting this axis. Show that the crystal has linear electro-optic coefficients of the form

$$\begin{pmatrix} 0 & 0 & r_{13} \\ 0 & 0 & r_{13} \\ 0 & 0 & r_{33} \\ 0 & r_{51} & 0 \\ r_{51} & 0 & 0 \\ 0 & 0 & 0 \end{pmatrix}.$$

Assuming linear effects only, obtain the form of the indicatrix when an electric field is applied in the direction of the sixfold rotational symmetry axis, denoted x_3.

Obtain an expression for the birefringence of light travelling through the crystal at right angles to the x_3 axis.

A crystal sample is cut as a parallelpiped with length 2 cm in the x_3 direction and with square cross section of side 3 mm. A sinusoidal voltage of amplitude 500 V is applied between two opposite side faces. Calculate the phase modulation for light of wavelength 3.39 μm.

(Data for CdSe at 3.39 μm: $n_O = 2.452$; $n_E = 2.417$; $r_{13} = 1.8 \times 10^{12}$ m V^{-1}; $r_{33} = 4.3 \times 10^{-12}$ m V^{-1}.)

7.3 Cadmium telluride has zinc-blende structure (crystal class $\bar{4}$3m). That is, it is cubic and has threefold rotational symmetry about the $\langle 111 \rangle$ directions, three fourfold inversion axes in the $\langle 100 \rangle$ directions, and six [110] mirror planes. Show that if an electric field E is established in the [111] direction, the indicatrix takes the form

$$\left(x_1^2 + x_2^2 + x_3^2 \right) / n_O^2 + 2r_{41} E \left(x_2 x_3 + x_3 x_1 + x_1 x_2 \right) / \sqrt{3} = 1.$$

Using a suitable transformation of axes, obtain the birefringence for light passing perpendicular to the [111] direction. Hence obtain the phase retardation produced for light of wavelength λ by a crystal of length L.

At 1.00 μm, the refractive index of CdTe is 2.84 and $r_{41} = 4.5 \times 10^{-12}$ m V^{-1}. A crystal of CdTe cut as a rectangular parallelpiped with $L = 8$ cm has a voltage V applied across its width of 0.4 cm. Calculate the value of V required to produce a phase retardation of π (i.e. to produce a half-wave plate).

7.4 Show than in an isotropic medium there are only two independent Kerr components s_{11} and s_{12} and that $s_{44} = (s_{11} - s_{12})/2$. All other components are either zero or equal to these components.

7.5 For a beam deflector, the aperture of the beam is set by diffraction and the angular width is $\theta_d = \lambda/d$ where d is the thickness of the device (figure 7.7). This determines the number of angular positions N that can be separately identified ($N = \theta/\theta_d$). Calculate the required electric field between the side faces for a KDP device of length L of 30 mm working at a wavelength of 633 nm to resolve ten points separately. (Use the data given in table 7.2.)

7.6 A crystal of KDP is used for second harmonic generation by phase matching at a fundamental frequency of 694 nm (ruby laser light). Calculate the matching angle measured from the optic axis ($n_E^\omega = 1.466$; $n_E^{2\omega} = 1.487$; $n_O^\omega = 1.506$; $n_O^{2\omega} = 1.534$).

7.7 A transverse electro-optic device can be constructed in which the light does not have to pass through coated end faces (figure 7.9). If the crystal is located between cross polarisers whose orientation is along the angular bisectors of axes Ox_1' and Ox_2' calculate the transmitted intensity of the light for an input intensity of I_0. Hence show that for a device whose length produces a static phase difference of 2π, the transmitted intensity is given by

$$I = \frac{1}{2} I_0 \sin^2 \left(\frac{\pi n_O^2 r_{63} L}{2\lambda d} V \right)$$

such that half-wavelength voltage will be

$$V_\pi = \frac{\lambda d}{n_O^2 r_{63} L}.$$

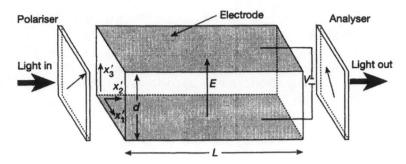

Figure 7.9 Transverse electro-optic device.

Calculate this voltage for a crystal of ADP of length 6.0 cm and square cross section of 9.0 mm² operating at a wavelength of 633 nm (see table 7.2).

8   Further Tensor Applications

8.1   Introduction

The topics included in this chapter are very varied in their nature. They illustrate further features of the application of tensors to physical properties. They are included in this separate chapter as none of the examples should be regarded as central to a basic understanding of the subject. Yet each example demonstrates an interesting point.

8.2   Thermal Expansion

When we consider heating of crystals, this process can give rise to expansions and sometimes contractions of different magnitudes in different directions. The expansions in three dimensions in a crystal constitute strain and the thermal expansion tensor is a second-rank tensor which relates the strain to the change in temperature. We considered the strain tensor in Chapter 4 and demonstrated its symmetrical second-rank nature. Measurement of change of dimensions of crystals with temperature in different directions can be made most easily by finding the variation of lattice spacings by crystallographic techniques. Having measured the strain for a number of different directions, the principal coefficients of expansion can be found. These will have values specific to particular crystals and the expansion tensor is a matter tensor as mentioned on p.56. The thermal expansion tensor is essentially a strain tensor which arises from the internal forces consequent on a change of temperature. This is in contrast with the strain tensor used to describe change of dimensions with load, where the tensor is a field tensor.

Let us assume that the temperature of our crystal increases uniformly by a small amount ΔT and that this produces strain components

$$\epsilon_{ij} = \alpha_{ij}\Delta T \tag{8.1}$$

where α_{ij} are components relating to the strain to the temperature rise in our particular crystal. $[\epsilon_{ij}]$ is the strain tensor and is symmetrical. Therefore $[\alpha_{ij}]$ must be a second-rank symmetrical tensor.

If we refer our tensors to their principal axes we obtain

$$\epsilon_1 = \alpha_1\Delta T$$
$$\epsilon_2 = \alpha_2\Delta T$$
$$\epsilon_3 = \alpha_3\Delta T$$

where α_1, α_2 and α_3 are the principal expansion coefficients. They are the components required to measure the thermal expansion of the crystal and the directions of the principal axes will be related to the directions of the crystallographic axes. Thus, if we write out the representation quadric for thermal expansion, referred to the principal axes, we obtain

$$\alpha_1 x_1^2 + \alpha_2 x_2^2 + \alpha_3 x_3^2 = 1. \tag{8.2}$$

The shape, as well as the orientation, of the quadric will be restricted by the crystal symmetry, i.e. by Neumann's principle (see p.31). Thermal expansion will not cancel any symmetry elements provided the crystal does not go through a phase transformation.

It is important to realise that equation (8.2) represents the crystal expansion but does not represent the shape of the expanded crystalline material. Suppose we take a spherical sample of crystal (we could take a spherical volume within an arbitrarily shaped crystal). After a temperature increase ΔT the crystal volume becomes ellipsoidal in shape with axes $(1 + \alpha_1\Delta T)$, $(1 + \alpha_2\Delta T)$ and $(1 + \alpha_3\Delta T)$. Only along the principal axes will the change be restricted to a change in length. Any line not parallel to a principal axis will exhibit a rotation relative to the system of axes.

The expansion coefficients α_1, α_2 and α_3 are usually all positive so that the representation quadric is ellipsoidal. However, a number of materials exhibit negative coefficients so that the thermal expansion quadric is a further example which can illustrate the various alternatives described on p.29.

The change in overall volume of a cube of side L for a 1 K temperature change will be given by

$$\delta V = L^3(1 + \alpha_1)(1 + \alpha_2)(1 + \alpha_3) - L^3$$
$$= V(\alpha_1 + \alpha_2 + \alpha_3)$$

to first order. Hence the volume expansivity, $\alpha_V = \delta V / V$, is given by

$$\alpha_V = (\alpha_1 + \alpha_2 + \alpha_3). \tag{8.3}$$

8.3   The Pyroelectric Effect

In the pyroelectric effect, a change in temperature leads to a release or movement of charge and the rate of spontaneous polarisation or change of electrical dipole moment, ΔP_i, is related to the temperature change ΔT by the pyroelectric coefficients p_i which are components of a vector which represents the crystal property pyroelectricity. ΔT is a small uniform temperature change within the crystal. Thus

$$\Delta P_i = p_i \Delta T. \tag{8.4}$$

The magnitude of the effect will differ according to whether the crystal is clamped during heating, so that it can change neither its shape not its size, or whether it is free to expand. When the pyroelectric effect is observed under the first condition it is called primary pyroelectricity and when it is observed under the second condition (free expansion) there is an additional effect called secondary pyroelectricity. No crystal showing a vectorial property can possess a centre of symmetry (something which is self-evident), so that the only crystals which can exhibit the pyroelectric effect are 1, 2, m, 2mm, 4, 4mm, 3, 3m, 6 and 6mm.

In the most general case, the triclinic class 1, the direction of p is not fixed with respect to the crystal structure. In the monoclinic classes, p must lie in the mirror plane for class m and parallel to the twofold rotational axis for class 2. In all the other cases there is a unique crystallographic axis to which p must lie parallel.

All pyroelectric materials are also piezoelectric (although the reverse is not true) and this effect will be considered in the next section.

8.4 Piezoelectricity

This is an example of a third-rank tensor. The direct piezoelectric effect is the development of an electric dipole moment P in a crystal when a stress σ is applied to the crystal. The magnitude of the electric moment is proportional to the stress, so we can write

$$P = d\sigma$$

where d is the piezoelectric modulus.

As we have seen, in general the stress can be represented by a second-rank tensor, and the polarisation can be in any direction and is a vector. So we may write

$$P_i = d_{ijk}\sigma_{jk} \tag{8.5}$$

and d_{ijk} are now the piezoelectric moduli.

A general third-rank tensor would require 27 components but σ_{jk} is symmetric in jk and hence d_{ijk} will be symmetric in jk also. This reduces the number of components to 18 and a triclinic crystal of point group 1 can have 18 independent components. A matrix notation can be used for the piezoelectric moduli in the same way as was used for stiffness and compliance in Chapter 4. The first subscript of the tensor notation remains the first subscript of the matrix notation and the second and third subscripts of the tensor notation become single matrix subscripts 1 to 6 following the pattern on p.58. Hence the piezoelectric moduli are represented by a 6×3 matrix and in subscript notation by

$$P_m = d_{mn}\sigma_n \tag{8.6}$$

where $m = 1, 2, 3$ and $n = 1, 2, \ldots, 6$.

As with compliance, it is necessary to introduce factors in the conversion from d_{ijk} to d_{mn}.

$$d_{ijk} = d_{mn} \qquad \text{for } n = 1, 2, 3$$
$$2d_{ijk} = d_{mn} \qquad \text{for } n = 4, 5, 6.$$

The matrix array for the piezoelectric moduli then takes the form

$$\begin{pmatrix} d_{11} & d_{12} & d_{13} & d_{14} & d_{15} & d_{16} \\ d_{21} & d_{22} & d_{23} & d_{24} & d_{25} & d_{26} \\ d_{31} & d_{32} & d_{33} & d_{34} & d_{35} & d_{36} \end{pmatrix}.$$

Remember that the d_{mn} do not transform as tensors, whereas the d_{ijk} do.

There is also the converse piezoelectric effect in which the application of an electric field E produces a strain within the crystal.

$$\epsilon_{jk} = d_{ijk} E_i. \tag{8.7}$$

Here the d_{ijk} are the same moduli as those in equation (7.3), a fact which can be proved thermodynamically. To change to the matrix form, the same conversions involving factors of $\frac{1}{2}$ are used in the conversion of ϵ_{jk} to ϵ_n as were used on p.59 for ϵ. This then fits with the use of the factors of $\frac{1}{2}$ to convert d_{ijk} to d_{mn}. Overall we have in matrix form

$$\epsilon_n = d_{mn} E_m \tag{8.8}$$

where $m = 1, 2, 3$ and $n = 1, 2, \ldots, 6$. As in the case of pyroelectricity, the piezoelectric effect can be measured under differing conditions, notably under constant stress or under constant strain. The principles of applying tensors to the effect remain the same in each case and the details will not be pursued further.

8.5   Photoelasticity

When a stress is applied to a crystal, not only can an electric moment be produced, but there can also be a change of dielectric constant and hence a change of refractive index. This change in the refractive index is called the photoelastic effect. The stress will be directional so an anaxial material will become doubly refracting and a uniaxial or biaxial crystal will show a change of components. The phenomenon has particular application to the representation of large solid objects such as machinery by small Perspex models. These models can be loaded in a similar manner to the larger original objects and changes of refractive index measured by viewing the models whilst they are positioned between crossed polarisers. This illustrates the effect in a non-crystalline material which is isotropic until the stress is applied.

The photoelastic tensor relates the change in the optical impermeability $\Delta \eta_{ij}$ to the strain tensor σ_{kl} and we write

$$\Delta \eta_{ij} = \Delta \left(1/n^2\right)_{ij}$$
$$= \Pi_{ijkl}\sigma_{kl}. \tag{8.9}$$

The Π_{ijkl} are the components of the strain-optic tensor and are dimensionless. Previously we had equation (5.3) for the indicatrix given by

$$x_1^2/n_1^2 + x_2^2/n_2^2 + x_3^2/n_3^2 = 1$$

where n_1, n_2 and n_3 are the principal values of the refractive index.

Alternatively, we can write the indicatrix equation in the form

$$\eta_1 x_1^2 + \eta_2 x_2^2 + \eta_3 x_3^2 = 1 \tag{8.10}$$

where we are using principal values of the optical impermeability η_{ij}.

More generally

$$\eta_{ij}x_i x_j = 1. \tag{8.11}$$

When the change of optical impermeability is included, this equality becomes

$$\left(\eta_{ij} + \Pi_{ijkl}\sigma_{kl}\right)x_i x_j = 1. \tag{8.12}$$

It is worth mentioning here acousto-optics, the interaction of optical waves with acoustic waves in a crystal. This can lead to many interesting effect and applications including diffraction of the optical waves. Propagation of the acoustic wave produces strain in the crystal and it is the periodic nature of the strain field which sets up the diffraction effects. Hence the strain components will be related to the direction of passage of the acoustic waves and the magnitude of any effects will then depend on the magnitudes of the components of the photoelastic tensor.

8.6   Incommensurate Modulated Structures

The patterns and crystal structures described in Chapter 1 all show regular repetition in any direction. They possess translational symmetry in two dimensions in the case of a pattern or in three dimensions in the case of a crystal. They are periodic patterns or periodic crystals.

However, it is possible to have crystal structures in which the periodicity is perturbed by a deviation of the atoms from a basic structure, but this deviation is itself periodic. The periodicity of this deviation or modulation may or may not be a simple multiple or fraction of the repeat distance of the basic lattice. If is a simple multiple or fraction, then a superlattice (sometimes called a superstructure) is established. There are many examples of superlattices. Silicon carbide can exist in many superlattice varieties. More recently, superlattices have become of relevance in semiconductor work where superlattices are produced from two or more semiconductors. For instance, m layers of gallium arsenide, GaAs, followed by n layers of aluminium arsenide, AlAs, repeated, will produce a superlattice $(GaAs)_m (AlAs)_n$ which repeats over m and n layers of GaAs and AlAs respectively. With crystal growth available by molecular beam epitaxy, many possible superlattice structures can be grown.

However, if the modulation is *not* a simple multiple or fraction then the modulation is described as incommensurate with the basic lattice. The overall structure can be described as an incommensurate modulated structure.

We can see how this comes about in one direction, the x-direction say. We start with a simple grid of atoms and then show them displaced as if by a transverse standing wave in the x direction but such that the deviation of the atoms is in the y direction. The wavelength of the wave is not a simple multiple of the basic lattice vector. Alternatively we could displace the atoms in the x direction as if we had set up a longitudinal standing wave, in which case the displacements are in the same direction as the modulation (figure 8.1).

Another way of having an incommensurate structure is to have intergrowth of two different crystal structures, each having its own lattice vectors. Figure 8.2 shows this and, so that this type of modulation can be compared with that in figure 8.1, two crystal

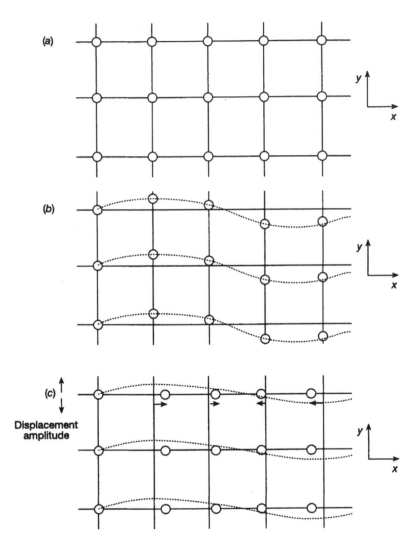

Figure 8.1 Formation of incommensurate modulated structures: (*a*) shows the basic lattice; (*b*) shows displacement from the basic lattice in the *y* direction but periodic in *x*; (*c*) shows similar modulation periodic in *x* but with displacement also in the *x* direction.

structures are chosen with their lattice vectors differing in the x direction only, but where the lattice parameter of one is not a simple multiple or fraction of the other. The resulting structure is regular throughout and constitutes a single thermodynamic phase.

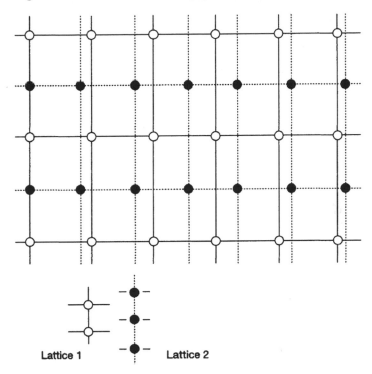

Figure 8.2 Incommensurate modulated structure consisting of two interpenetrating crystal structures.

We can go a further stage to consider quasicrystals. These can be illustrated in two dimensions by so-called Penrose tiling. In the case illustrated in figure 8.3 the two-dimensional pattern is made up of two rhombohedra, one of each type having been shaded in. However, the arrangement fails to fit any regular modulation. Three-dimensional quasicrystals have been discovered and investigated, for instance crystals of Al_6CuLi_3, and non-crystallographic point symmetry associated with them.

What is needed is a method of classifying and hence distinguishing the possible incommensurate phases from each

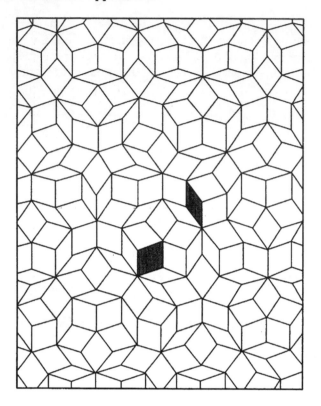

Figure 8.3 A section of two-dimensional Penrose tiling consisting of two rhombohedrals (shaded) repeated with no overall periodicity in any direction.

other and from crystals merely periodic as for a simple lattice. One way of doing this is to consider variation of density with position within the lattice. In the case of a crystal made up of repeat cells, or building blocks, with edges of lengths a, b and c, the density will repeat for all displacements a, b and c, and any combination and any number of these displacements, i.e. for any total displacement $L = n_1a + n_2b + n_3c$, where n_1, n_2 and n_3 are integers. Hence we have a density function $\rho(r + L)$ which is equal to $\rho(r)$, the variation of density with position·in the originating cell.

For an incommensurately modulated crystal we now superimpose a modulation which we might represent as sinusoidal in the x direction as represented in figure 8.1(c) such that density varies as $\rho(r + L)\sin(2\pi x)$ where the periodicity is not a multiple of a†. It is a relatively easy concept, but rather more difficult to picture mathematically, that this is a repeat distance that effectively can be represented as a fourth direction within a four-dimensional space. If one goes on to consider a more general modulation taking place in any arbitrary direction relative to the original crystal lattice, then one can represent the periodicity with three parameters and extend to six-dimensional space. The extra dimension or dimensions can be incorporated within the tensor formulation described in Chapter 2 but, whereas p and q are there assumed to vary from 1 to 3 to represent three Cartesian directions, here the variation can go beyond three to represent the additional modulation. However, a full interpretation becomes complicated when considering the type of dimension in which variation is occurring and the type of physical effect under consideration. So far considerable work has been done in obtaining space group representations for quasiperiodic structures (this text has avoided discussion of group theory) and the extension of the work to tensor properties depends both on extending the theoretical calculations and carrying out experimental work on incommensurately modulated structures. Texts for students on incommensurate structures are not yet readily available but various review articles have appeared (for instance, Janssen and Janner 1987, Van Smaalen 1995).

Incommensurate structures have been particularly discovered in certain high temperature superconductors and related materials. These materials exhibit extra satellite reflection within their electron diffraction patterns that can only come from incommensurate modulation within their structures. Such modulation is usually found in materials which go through solid–solid phase transitions as temperature is varied, a feature which is characteristic of high temperature superconductors. The high temperature superconductors usually possess an underlying perovskite arrangement, the basic arrangement exhibited by sodium niobate $NaNbO_3$. Here the niobium atoms sit at the centres of octahedra which have oxygen atoms at their vertices. The sodium atoms sit at the centres of the spaces between the octahedra. In the different materials, the octahedra can be arranged in a variety of patterns. Even in the relatively straightforward case of sodium niobate, the

---

† Strictly for this discussion we should work with reciprocal lattice vectors and reciprocal lattice space, but this has been avoided within this textbook.

compound exhibits at lower temperatures a superlattice modulation involving four units, another modification involving six units, and an incommensurate phase with repeat displacements slightly more than six units. Other types of materials where such structures exist and where solid–solid phase transitions occur are ferromagnetics and ferroelectrics. The topic of incommensurability and its classification is likely to be a fruitful area of materials research and application in the future.

8.7   Some Concluding Comments

In a book of this size, many effects which can be described by tensors have been omitted but the techniques used here can be extended to these other effects. For instance in Chapter 6 we looked at the effect of a magnetic field on electrical and thermal conductivity and in Chapter 7 we looked at the effect of an electric field on the passage of optical waves through a crystal. We could go on to consider the effect of a magnetic field on the propagation of light through a crystal. A magnetic field can, in general, alter the dielectric tensor and produce a whole range of effects. For instance, one can define a magnetogyration coefficient for the rotation of light in a crystal produced as a result of an applied magnetic field. Such an effect is called the Faraday effect.

A very different type of effect which has been omitted is the thermoelectric effect. To understand thermoelectricity it is necessary to establish a considerable background terminology which was felt to be inappropriate in understanding tensor applications. Appendix 3 gives a selected list of properties for which tensors can be used.

Problems

8.1   Show by symmetry arguments that the matrix array for KDP (crystal class $\bar{4}2m$) for the piezoelectric case is given by

$$\begin{pmatrix} 0 & 0 & 0 & d_{14} & 0 & 0 \\ 0 & 0 & 0 & 0 & d_{14} & 0 \\ 0 & 0 & 0 & 0 & 0 & d_{36} \end{pmatrix}.$$

8.2 Consider the compound $Pb_2(Sc, Ta)O_6$ which has a modified cubic ABO_3 perovskite structure. The B atom is replaced by equal amounts of Sc^{3+} and Ta^{5+}. Show that this is compatible with a doubling of the cell in each direction. Hence indicate why a superlattice structure will be observed if the Sc^{3+} and Ta^{5+} elastically scatter x-rays or similar radiation by different amounts.

8.3 The Faraday effect is a rotation of the plane of polarisation of light with distance when a material is placed in a magnetic field. The specific rotation per unit length is given by VB where V is the Verdet constant and B is the magnetic field. Compare the lengths of sample required for 2.5 degrees rotation for fluorite, diamond and sodium chloride using a magnetic field of 0.5 tesla. (V/deg T^{-1} mm^{-1}: fluorite 0.015; diamond, 0.20; sodium chloride, 0.60.)

Appendix 1

Number of Independent Components

Table A1 The maximum number of independent coefficients occurring for different rank tensors for the 32 crystal classes.

International notation and crystal system	Second-rank symmetrical	Second-rank non-symmetrical	Third-rank	Fourth-rank symmetrical	Fourth-rank optical	Axial second-rank
1 triclinic	6	9	18	21	36	6
$\bar{1}$	6	9	0	21	36	0
2 monoclinic	4	5	8	13	20	4
m	4	5	10	13	20	2
2/m	4	5	0	13	20	0
222 orthorhombic	3	3	3	9	12	3
mm2	3	3	5	9	12	1
mmm	3	3	0	9	12	0
4 tetragonal	2	3	4	7	10	2
$\bar{4}$	2	3	4	7	10	2
4/m	2	3	0	7	10	0
422	2	2	1	6	7	2
4mm	2	2	3	6	7	0
$\bar{4}$2m	2	2	2	6	7	1
4/mmm	2	2	0	6	7	0

Table A1 (Continued)

International notation and crystal system	Second-rank symmetrical	Second-rank non-symmetrical	Third-rank	Fourth-rank symmetrical	Fourth-rank optical	Axial second-rank
3 trigonal	2	3	6	7	12	2
$\bar{3}$ (rhombo-	2	3	0	7	12	0
32 hedral)	2	2	2	6	8	2
3m	2	2	4	6	8	0
$\bar{3}$m	2	2	0	6	8	0
6 hexagonal	2	3	4	5	8	2
$\bar{6}$	2	3	2	5	8	0
6/m	2	3	0	5	8	0
622	2	2	1	5	6	2
6mm	2	2	3	5	6	0
$\bar{6}$m2	2	2	1	5	6	0
6/mmm	2	2	0	5	6	0
23 cubic	1	1	1	3	4	1
m3	1	1	0	3	4	0
432	1	1	0	3	3	1
$\bar{4}$3m	1	1	1	3	3	0
m3m	1	1	0	3	3	0
Isotropic without centre of symmetry	1	1	0	2	2	1
Isotropic with centre of symmetry	1	1	0	2	2	0
Typical examples	Resistivity	Thermo electricity, Hall effect	Piezo-electricity, electro-optic effect (linear)	Elastic compliance	Elasto-optic, piezo-optic effects	Optical activity

Appendix 2

Diagonalisation of a Tensor

We have assumed throughout this book that a tensor for which we can use the representation

$$S_{pq}x_p x_q = 1$$

can be simplified to

$$S_1 x_1^2 + S_2 x_2^2 + S_3 x_3^2 = 1 \qquad (2.14)$$

by a change of axes, but we have not proved this or shown the magnitudes of the rotations required.

Equation (2.14) is an equation in which x_1, x_2 and x_3 are referred to the principal axes. At the intersections of these principal axes with the representation quadric, the normals to the quadric will be parallel to the corresponding radius vectors (i.e. parallel to the corresponding principal axes).

Take P as such an intersection on the quadric and $OP = X_i$ where i can take values 1, 2, 3. Then by the radius normal property on p.40

$$S_{ij}X_j = \lambda X_i$$

where λ is a constant. This equality represents three homogeneous equations whose determinant must be zero, i.e.

$$\begin{bmatrix} S_{11} - \lambda & S_{12} & S_{13} \\ S_{12} & S_{22} - \lambda & S_{23} \\ S_{13} & S_{23} & S_{33} - \lambda \end{bmatrix} = 0.$$

There are three roots for λ giving three non-zero solutions which define the directions of the principal axes. Numerical

solution is not easy and is best achieved by successive approx-
imation. (Solution by computer is clearly advantageous.)
    If the quadric is referred already to its principal axes, then

$$\begin{bmatrix} S_1 - \lambda & 0 & 0 \\ 0 & S_2 - \lambda & 0 \\ 0 & 0 & S_3 - \lambda \end{bmatrix} = 0$$

$$(S_1 - \lambda)(S_2 - \lambda)(S_3 - \lambda) = 0$$

and the roots arc S_1, S_2 and S_3.

Appendix 3

Table of Tensor Properties

Type of tensor	Property	Defining equation†	Refer
First-rank relating scalar and vector	Pyroelectricity	$\Delta P_i = p_i \Delta T$	p.135
	Polarisation by hydrostatic pressure		
Second-rank symmetrical relating two vectors	Thermal conductivity	$h_i = -k_{ij}(\partial T/\partial x_j)$	p.36
	Thermal resistivity	$\partial T/\partial x_i = r_{ij}h_j$	p.36
	Electrical conductivity	$j_i = \sigma_{ij}E_j$	p.41
	Electrical resistivity	$E_i = \rho_{ij}j_j$	p.42
	Diffusion	$J_i = D_{ij}(\partial C/\partial x_j)$	p.43
	Permittivity	$D_i = K_{ij}E_j$	p.72
	Magnetic permeability		

† See pp.xi–xiii for list of symbols.

Type of tensor	Property	Defining equation†	Refer
Second-rank non-symmetrical relating two vectors	Thermoelectricity		
Second-rank symmetrical relating scalar and second-rank	Thermal expansion	$\varepsilon_{ij} = \alpha_{ij}\Delta T$	p.134
	Strain by hydrostatic pressure		
Second-rank non-symmetrical relating scalar and second-rank	Peltier		
Axial second-rank relating pseudo-scalar with direction	Optical activity	$G = g_{ij}l_il_j$	p.96
Third-rank vector and second-rank	Direct piezoelectric	$P_i = d_{ijk}\sigma_{jk}$	p.136
	Converse piezoelectric	$\varepsilon_{jk} = d_{ijk}E_i$	p.137
	Linear electro-optic (Pockels)	$\Delta\eta_{ij} = r_{ijk}E_k$	p.106
Third-rank axial relating axial vector to antisymmetrical second-rank (or second-rank antisymmetrical)	Hall effect	$a_{ij} = \rho_{ijk}B_k$	p.100

† See pp.xi–xiii for list of symbols.

Type of tensor	Property	Defining equation†	Refer
Fourth-rank relating two second-ranks	Elastic compliance	$\varepsilon_{ij} = s_{ijkl}\sigma_{kl}$	p.58
	Elastic stiffness	$\sigma_{ij} = c_{ijkl}\varepsilon_{kl}$	p.58
	Photoelasticity	$\Delta\eta_{ij} = \Pi_{ijkl}\sigma_{kl}$	p.138
	Quadratic electro-optic (Kerr)	$\Delta\eta_{ij} = s_{ijkl}E_k E_l$	p.106
	Magneto-resistance	$E_i = \rho_{ijkl}J_j B_k B_l$ (usually $k = l$)	p.100

Field tensors

Second-rank relating two vectors	Stress		
	Strain	$\Delta u_i = [\varepsilon_{ij}]\delta x_j$	p.54

† See pp.xi–xiii for list of symbols.

References and Further Reading

Bhagavantam S 1966 *Crystal Symmetry and Physical Properties* (London: Academic)
Similar material is covered to that in this book. There is less emphasis on optical properties and there is reference to group theory which has been excluded here. It is a more theoretical text with few diagrams and a lack of worked examples. However, there are extensive lists of tensor components for different crystal classes.

Butcher P N and Cotter D 1990 *The Elements of Nonlinear Optics* (Cambridge: Cambridge University Press)
A review of nonlinear optics at a more advanced level than material in this book. It includes a chapter on symmetry properties and susceptibility and *d*-tensor tables.

Burns G and Glazer A M 1990 *Space Groups for Solid State Scientists* 2nd edn (San Diego, CA: Academic)
A comprehensive text on crystal structures including superstructures and incommensurately modulated materials.

Huard S 1977 *Polarization of Light* (Chichester: Wiley, Paris: Masson)
A detailed account of optical effects including optoelectronic devices.

Janssen T and Janner A 1987 Incommensurability in crystals *Adv. Phys.* **36** 519–624
A review of the topic.

Kelly A and Groves G W 1970 *Crystallography and Crystal Defects* (London: Longman)
A good introduction to crystallography with a tensor section which mainly concerns stress, strain and elasticity.

Lipson S G and Lipson H 1995 *Optical Physics* (Cambridge: Cambridge University Press)
A standard undergraduate text on optics which includes a useful section on crystal optics with extensive use of tensors.

McKie D and McKie C 1974 *Crystalline Solids* (New York: Wiley)
An extensive text on crystallography including two chapters on the use of tensors. Diffusion, thermal expansion and crystal optics are discussed in particular.

Nye J F 1985 *Physical Properties of Crystals* (Oxford: Oxford University Press)
This remains, since its 1957 first edition, the standard text on the representation of physical properties of crystals by tensors, and frequent reference has been made to it. The book covers all the main topics referred to here, usually to a more advanced level, but is limited in its worked examples.

Phillips F C 1971 *An Introduction to Crystallography* 4th edn (Edinburgh: Oliver and Boyd)
A good approach to crystallography with descriptions of all the crystal classes.

Popov S, Svirko Y and Zheludev N 1998 *Encyclopedia of Material Tensors on CD* (Chichester: Wiley)
A database of 130 000 material tensors of ranks two to seven.

Popov S V, Svirko Y P and Zheludev N I 1995 *Susceptibility Tensors or Nonlinear Optics* (Bristol: Institute of Physics)
Book and CD-Rom covering ranks two to five.

Van Smaalen S 1995 Incommensurate crystal Structures *Cryst. Rev.* 4 79–202
A review of the topic.

Woolfson M M 1970 *An Introduction to X-ray Crystallography* (Cambridge: Cambridge University Press)

This constitutes a standard introduction to crystallography and is probably more attractive to physicists than some other crystallography texts.

Wooster W A 1973 *Tensors and Group Theory for the Physical Properties of Crystals* (Oxford: Clarendon)
The two parts of this book are very dissimilar. The first half has similar coverage to this book and includes worked examples. The second half uses group theory and is more theoretically oriented; it covers different ground from this book.

Yariv A and Yeh P 1984 *Optical Waves in Crystals* (New York: Wiley)
An advanced text on crystal optics with particular coverage of electro-optics and acousto-optics including examples. It includes tables of coefficients, both tensor components and coefficients for actual materials.

Answers to Problems

Chapter 1

1.1 oblique system: class 1; α, β, δ, ε
                      class 2; θ, $\int$
     rectangular system: class m; > (horizontal)
                                Π (vertical)
                     class 2mm; Φ, ∞

1.2 See figure P

1.3 (*a*) (i) ($\bar{1}$10)   (ii) (110)   (iii) ($\bar{1}$11)
             (iv) ($\bar{1}$10)   (v) ($\bar{1}$00)
             (vi) (010)   (or negative equivalents)
     (*b*) (vii) [012]   (viii) [221]   (ix) [$\bar{1}\bar{1}$2]   (x) [1$\bar{1}$2]

1.4 (i) 90°   (ii) 54.7°   (iii) 26.6°   (iv) 18.4°

Chapter 2

2.1

$$\begin{bmatrix} 16 & 0 & 0 \\ 0 & 16 & 0 \\ 0 & 0 & 4 \end{bmatrix}$$

(a)

(b)

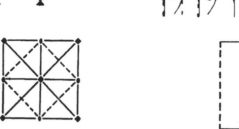

(c)

(d)

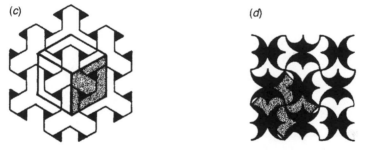

Figure P Solutions to Problem 1.2. (*a*) 4, (*b*) 11g (i.e. glide only), (*c*) 6, (*d*) 4gm.

Chapter 3

3.1 87.5 W
3.2 $1.32 \times 10^{-6}\,\Omega\,\text{m}$ 4.9°
3.3 $1.93 \times 10^{-7}\,\Omega\,\text{m}$
3.4 $2.1\,\text{W}\,\text{cm}^{-2}$ 30°

Chapter 4

4.1

$$\begin{bmatrix} 12 & 6 & 4 \\ 6 & 7 & 5 \\ 4 & 5 & 2 \end{bmatrix} + \begin{bmatrix} 0 & -2 & -1 \\ 2 & 0 & 1 \\ 1 & -1 & 0 \end{bmatrix}$$

4.2

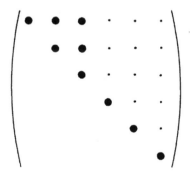

4.3

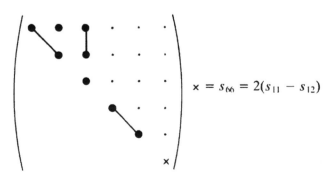

$$\times = s_{66} = 2(s_{11} - s_{12})$$

Chapter 5

5.1 2.6 μm

5.2 (i) 1.84 μm   (ii) 3.58 μm

5.4 50° to the k_1 axis

Chapter 6

6.1 (a)
$$\begin{bmatrix} g_{11} & 0 & 0 \\ & g_{22} & 0 \\ & & g_{33} \end{bmatrix}$$
(b)
$$\begin{bmatrix} 0 & g_{12} & 0 \\ & 0 & 0 \\ & & 0 \end{bmatrix}$$

6.2

$\bullet = \rho_{231}$ $\circ = \rho_{321}$

Chapter 7

7.2 218°

7.3 $(x_1^2 + x_2^2 + x_3^2)/n_O^2 + 2r_{41}E(x_2x_3 + x_3x_1 + x_1x_2)/\sqrt{3} = 1$

Axes transformations of form:

	x_1'	x_2'	x_3'	Σl^2
x_1	$1/\sqrt{3}$	$-1/\sqrt{2}$	$-1/\sqrt{6}$	1
x_2	$1/\sqrt{3}$	$1/\sqrt{2}$	$-1/\sqrt{6}$	1
x_3	$1/\sqrt{3}$	0	$2/\sqrt{6}$	1
Σl^2	1	1	1	

$\sqrt{3}\pi n_O^3 r_{41} EL/\lambda$ 280 V

7.5 5900 V mm^{-1}
7.6 2.3 kV
7.7 $50.4°$

Chapter 8

8.3   3.3 cm, 2.5 cm, 8.3 mm

Index

Printed and bound by CPI Group (UK) Ltd, Croydon, CR0 4YY

17/10/2024

01775686-0020